Theory and Applications of Computability

In cooperation with the association Computability in Europe

Series Editors

Prof. P. Bonizzoni
Università Degli Studi di Milano-Bicocca
Dipartimento di Informatica Sistemistica e Comunicazione (DISCo)
20126 Milan
Italy
bonizzoni@disco.unimib.it

Prof. V. Brattka
University of Cape Town
Department of Mathematics and Applied Mathematics
Rondebosch 7701
South Africa
vasco.brattka@uct.ac.za

Prof. S.B. Cooper
University of Leeds
Department of Pure Mathematics
Leeds LS2 9JT
UK
s.b.cooper@leeds.ac.uk

Prof. E. Mayordomo
Universidad de Zaragoza
Departamento de Informática e Ingeniería de Sistemas
E-50018 Zaragoza
Spain
elvira@unizar.es

For further volumes:
http://www.springer.com/series/8819

Books published in this series will be of interest to the research community and graduate students, with a unique focus on issues of computability. The perspective of the series is multidisciplinary, recapturing the spirit of Turing by linking theoretical and real-world concerns from computer science, mathematics, biology, physics, and the philosophy of science.

The series includes research monographs, advanced and graduate texts, and books that offer an original and informative view of computability and computational paradigms.

Douglas S. Bridges • Luminiţa Simona Vîţă

Apartness and Uniformity

A Constructive Development

 Springer

Douglas S. Bridges
Department of Mathematics
and Statistics
University of Canterbury
Private Bag 4800
Christchurch
New Zealand
d.bridges@math.canterbury.ac.nz

Luminiţa Simona Vîţă
Department of Mathematics
and Statistics
University of Canterbury
Private Bag 4800
Christchurch
New Zealand
simona.vita@canterbury.ac.nz

ISSN 2190-619X   e-ISSN 2190-6203
ISBN 978-3-642-26996-7   ISBN 978-3-642-22415-7 (eBook)
DOI 10.1007/978-3-642-22415-7
Springer Heidelberg Dordrecht London New York

ACM Computing Classification (1998): F.1, F.4

AMS Codes: 03-XX, 68-XX

Cover design: deblik, Berlin

Printed on acid-free paper

Springer is part of Springer Science+Business Media (www.springer.com)

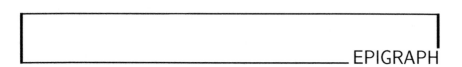

EPIGRAPH

I may prefer my ideals. The fact that they may not survive is not a reason for not fighting for them. Schumpeter rightly said that people who believe that ideals have to be absolute are idolatrous barbarians. Civilization means that you must allow the possibility of change without ceasing to be totally dedicated to—and ready to die for—your ideals so long as you believe in them.

Isaiah Berlin[1]

---

[1]From *Conversations with Isaiah Berlin* (2nd ed.), by Ramin Jahanbegloo, Halban Publishers Ltd, London, 2007.

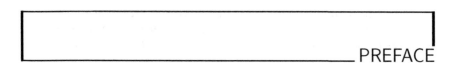

PREFACE

The problem of finding a suitable constructive framework for general topology is important and elusive.

Errett Bishop[2]

In the mid-1970s, Douglas Bridges came across the article [35], advocating the use of nearness in the teaching of first courses in analysis. This led him to consider a constructive theory based on proximity—or, rather, the constructively more appropriate opposite notion of apartness—as a possible alternative approach to topology. Not, at that stage, being sufficiently mathematically experienced to make significant progress with this idea, he put it on one side until February 2000, when he and Luminiţa Vîţă began the project on apartness spaces that is discussed in the present monograph. Clearly, two heads were better than one, as we were able to initiate a project that, in the intervening years, has grown substantially, with contributions from mathematicians in several countries and continents.

What do we mean by *constructive* in this context, and why might a constructive approach to topology be interesting or significant? The answer to the first question is simple: by *constructive mathematics* we mean, roughly, mathematics with intuitionistic logic and a set- or type-theoretic foundation that precludes the derivation of the law of excluded middle; in other words, we mean mathematics carried out in the style of the late Errett Bishop [9]. Every proof that is constructive in this sense embodies an algorithm that can be, and in several cases has been, extracted and then implemented; further, the original proof shows that the implementable algorithm (if correctly extracted)

---

meets its specifications. One other important feature of Bishop's constructive mathematics is that it is completely consistent with classical mathematics—mathematics using classical logic.

In regard to the second question, we recall Bishop's deflationary remark [9] (page 63):

> Very little is left of general topology after that vehicle of classical mathematics has been taken apart and reassembled constructively. With some regret, plus a large measure of relief, we see this flamboyant engine collapse to constructive size.

At the time he wrote [9], Bishop firmly believed that computation in analysis required uniform continuity—normally on compact sets—rather than the weaker notion of pointwise (let alone sequential) continuity:

> The concept of a pointwise continuous function is not relevant. A continuous function [on the real line] is one that is uniformly continuous on compact intervals. ([9], pages ix–x)

Since the theorem asserting the uniform continuity of every pointwise continuous, real-valued mapping on the closed interval $[0, 1]$ cannot be proved within Bishop-style constructive mathematics,[3] general topology, essentially a pointwise matter, would therefore have little constructive relevance; what was needed was a framework (such as metric-space theory) in which *uniform* notions could be expressed.

Bishop's remark, taken with his suggestion, in Appendix A to [9], that constructive theories like that of distributions might rely on ad hoc topological notions, may have served to deflect attention away from the problem of finding a suitable constructive framework for general topology. The advantage of solving this problem would presumably be that we would have a general, and hence almost certainly clarifying, framework for topological matters. Moreover, work of Ishihara [55] and others, 20 years after the publication of [9], has shown that even sequential continuity has an important role in Bishop-style mathematics; so there is definitely a place for a constructive theory of abstract spatial relationships that covers such highly non-uniform types of continuity.

One can certainly tackle topology by simply constructivising the classical development of the subject from the usual three axioms about open sets.[4] However, the theory of apartness that we present in this book does much more, by encompassing both point-set topology and the theory of uniform spaces. This, we believe, enables us to see those subjects in a clearer light.

We now outline the contents of the three chapters in our book, which begins with one that introduces informal constructive (intuitionistic) logic and set

---

[3]The theorem is provable in one model of Bishop-style mathematics—namely, intuitionistic mathematics—but is false in the model in which everything is interpreted recursively.

[4]This has been done by Grayson [46, 47], Troelstra [87], and Waaldijk [91], the last two working within Brouwer's full intuitionistic framework.

theory, as well as—very briefly—the basic notions and notations for metric and topological spaces.

In Chapter 2 we introduce axioms for a point-set apartness, a more computationally informative notion than nearness.[5] We then explore some of the elementary consequences of our axioms, and examine the associated topology and various types of continuity of mappings. One interesting feature of our constructive approach is that intuitionistic logic reveals distinctions that are invisible to the classical eye. For example, we present three natural notions of continuity for maps between point-set apartness spaces, notions that coalesce classically but are quite distinct constructively.

After dealing with continuity, the chapter continues with a discussion of convergence for nets in an apartness space, followed by a study of the apartness structure on the product of two apartness spaces. It ends with some remarks about impredicativity and how that phenomenon might be avoided.

The theory really gets interesting—and a lot harder—when, in Chapter 3, we consider apartness between subsets. The five axioms on which this part of the theory is based enable us to bring most of the work on point-set apartness across to the set-set environment. Every set-set apartness gives rise to a point-set apartness in the obvious way: a point x is apart from a set S if the singleton subset $\{x\}$ is apart from S.

The canonical example of a set-set apartness arises from a quasi-uniform structure (which, if a certain symmetry condition holds, becomes a uniform structure). Having introduced both apartness and quasi-uniform spaces, we are well placed to examine the connection between, on the one hand, the uniform continuity of a mapping $f : X \to Y$ between quasi-uniform spaces and, on the other, strong continuity, in which if the two subsets of Y are apart, then their pre-images under f are apart in X. It is straightforward to show that uniform continuity entails strong continuity; but the converse direction of entailment is much harder to deal with. Similar difficulties are found in our discussion of the relation between Cauchy nets in the usual uniform-space sense, and nets that have the property of *total Cauchyness* relative to the apartness structure associated with a given uniform structure. (The corresponding property of *total completeness* of an apartness space X, whereby every totally Cauchy sequence in X has a limit in X, is classically equivalent to the compactness of the underlying apartness topology on X.)

The difficulties, alluded to above, in the proofs of certain results are unified by our introduction of a general proof technique which has several applications in the theory and which resembles—but is considerably more complex than—a well-known technique introduced into constructive analysis by Ishihara [55]. Not only does our technique apply to the study of strong and uniform continuity, but also it leads to a neat proof that, under certain conditions, the uniform

---

[5]Actually, our fundamental notion is that of a *pre-apartness*. An apartness is a pre-apartness with an extra disjunction property that enables us to split arguments into cases.

convergence of a sequence of mappings between uniform spaces is equivalent to their being convergent in an apartness-space sense. (In order to provide a sound framework for the study of convergence of functions, we introduce structures, constructively weaker than apartness and quasi-uniform ones, on the space Y^X of mappings from a set X into an apartness space Y.)

After a discussion of totally Cauchy nets, the chapter continues by introducing notions of *locatedness* that lift that fundamental property of subsets of a metric space into the wider context of a uniform space. In particular, we connect locatedness and another important property, *total boundedness* (one that lifts more easily from the metric to the uniform context). We then deal with product apartness spaces, paying, as in the case of point-set apartness, particular attention to the flow of properties, such as total completeness, between a product apartness space and its two "factors".

In the next section of the chapter we discuss a form of connectedness for apartness spaces. Then we introduce a special structure, denoted by $\mathcal{B}_w$, on an apartness space X. We use this to discuss uniform structures that are compatible with a given apartness structure on a set. Although we cannot generally prove that there is such a uniform structure, we can show that there is at most one totally bounded uniform structure of this sort, which is then the smallest compatible uniform structure. In addition, the properties of $\mathcal{B}_w$ lead us to a notion of *nearness* between subsets of an apartness space. If the apartness space has a strong separation property, then the point-set restriction of this nearness coincides with the notion of nearness associated, in Chapter 2, with the point-set apartness on X. Finally, we discuss Diener's alternative approach to compactness in apartness spaces, based on his notions of *neat locatedness* and *neat compactness*.

There are other, successful and important, constructive approaches to topology that we should mention: the ones based on the notions of *frame*, *locale*, and *formal topology* [60, 61, 79, 80, 89]. Some work has been done towards clarifying the relation between these "point-free" approaches and our theory of apartness spaces [76]; this work is the subject of the Postlude at the end of the book. We believe that constructive mathematics has room for both the theory of apartness and the point-free approaches to topology, and that these should be regarded as equally valid and viable.

Christchurch and Wellington, New Zealand   *Douglas Bridges*
24 January 2011   *Luminiţa Simona Vîţă*

Acknowledgments

We gratefully acknowledge the support of the following individuals and funding bodies at various stages in the writing of this book:

- ↪ The Deutscher Akademischer Austausch Dienst (DAAD) for supporting Bridges for one year as a Gastprofessor at Ludwig-Maximilians-Universität, München (LMU).

- ↪ Otto Forster, Peter Schuster, and Helmut Schwichtenberg, our hosts at LMU in 2003 and on several other occasions.

- ↪ The New Zealand Foundation of Research, Science and Technology for Vîţă's Postdoctoral Fellowship from 2002–2005 and for supporting her extended visit to LMU to enable us to work together on the book.

- ↪ The Royal Society of New Zealand, for supporting Bridges's research in 2006–2009 by an award from its Marsden Fund.

- ↪ The New Zealand Ministry of Research, Science and Technology, for supporting the authors by counterpart funding for the EU IRSES project entitled *Constructive Mathematics: Proof and Computation* (project no. 230822), involving LMU; the Universities of Canterbury, Padova, and Uppsala; and the Japan Advanced Institute of Science and Technology (Ishikawa).

- ↪ The Department of Mathematics and Statistics at the University of Canterbury, for enabling us to work together in Christchurch and Wellington on several occasions in 2007–2010.

- ↪ The various mathematicians who have joined us to work on the apartness-space project since it began in 2000.

- ↪ Hannes Diener, Anton Hedin, Matt Hendtlass, Erik Palmgren, and Thomas Steinke, for reading our drafts, giving us valuable suggestions for improving them, and, in Matt's case, creating our diagrams.

It is a pleasure once again to thank our friends Imola and Atila Zsigmond, and Helmut and Eva Pellinger, for being such excellent hosts during our time in Munich.

Finally, we thank our families for their continuing love and support.

CONTENTS

1

THE CONSTRUCTIVE FRAMEWORK

He who has not first laid his foundations may be able with great ability to lay them afterwards, but they will be laid with trouble to the architect and danger to the building.

Niccolò Machiavelli

Synopsis

We begin by introducing the constructive framework within which our theory lies. Section 1.1 deals with informal intuitionistic logic and how its use affects even the elementary theory of real numbers. In Section 1.2 we describe the naive set theory needed in the book; and in Section 1.3 we give the briefest outline of the basic notions associated with metric and topological spaces.

1.1 Constructivity and Logic

One hundred and two years ago, in his talk on the *Theory of Enchainment* at the International Congress of Mathematicians (Rome, 1908), F. Riesz proposed using a primitive notion of nearness, or proximity, as the basis for investigating abstract spatial relationships between sets [78]. That proposal lay in abeyance until Efremovič, in the mid 1930s, created the axiomatic theory of proximity spaces (whose publication as [41] was delayed until 1951), a theory that has established itself as a significant, productive field of classical topological research.

1

In this monograph we present a theory partly inspired by the classical one of proximity, but distinguished from it in two significant aspects: first, our theory is based on the notion of apartness (between points and sets, and, later, between sets and sets), which is computationally more informative than that of proximity; secondly—and it is this second aspect that dictates our emphasis on apartness—we use intuitionistic logic, rather than classical logic, throughout.

Why do we make this logical shift? We do so in order to make our presentation fully constructive/computational, in the sense and style of Errett Bishop [9, 12, 24, 29]. We could have produced a computational theory of apartness using classical logic, but this would have committed us to a framework, such as recursive function theory, which enabled us to distinguish constructive objects and processes from nonconstructive ones. Instead, by using intuitionistic logic, we do not need to restrict ourselves to a recursive framework: intuitionistic logic (plus an appropriate set- or type-theoretic foundation such as those found in [2, 43, 70, 72]) automatically excludes nonconstructive arguments. Every argument that uses intuitionistic logic and does not lapse by admitting purely classical principles such as the **law of excluded middle** (**LEM**),

$$P \text{ or } (\text{not } P),$$

or certain weaker forms thereof such as the **weak limited principle of omniscience** (**WLPO**),

> For each binary sequence, either all the terms equal 0 or else it is impossible that all the terms equal 0,

will be constructive. In particular, if we give an intuitionistic-logic-based proof of the existence of a mathematical object x with a certain property $P(x)$, then we can extract from our proof an algorithm that enables us to construct (compute) an object ξ and then to prove that $P(\xi)$ holds. The extraction of algorithms from such proofs has actually been carried out by computer science research groups in several locations across the globe [36, 49, 70, 84].

It is important to appreciate from the outset that *the Bishop-style constructive mathematics* (**BISH**) *which forms the framework for this book is entirely consistent with classical mathematics*. Every theorem and proof in the former is valid within the latter. Indeed, we can regard classical mathematics as Bishop-style mathematics plus the law of excluded middle.

The intuitionistic/constructive interpretations of the various connectives

$$\vee \text{ (or)}, \quad \wedge \text{ (and)}, \quad \Rightarrow \text{ (implies)}, \quad \neg \text{ (not)}$$

and quantifiers

$$\exists \text{ (there exists)}, \quad \forall \text{ (for all/each)}$$

are as follows:

▷ $P \vee Q$: either we have a proof of P or we have a proof of Q.

▷ $P \wedge Q$: we have a proof of P and a proof of Q.

▷ $P \Rightarrow Q$: by means of an algorithm—that is, a finite, computational procedure—we can convert any proof of P into a proof of Q.

▷ $\neg P$: assuming P, we can derive a contradiction (such as $0 = 1$); equivalently, we can prove $(P \Rightarrow (0 = 1))$.

▷ $\exists_{x \in A} P(x)$: we have an algorithm which computes an object x and determines that it satisfies the conditions for membership (denoted, as usual, by the $\in$ symbol) of the set A, and an algorithm which demonstrates that $P(x)$ holds.

▷ $\forall_{x \in A} P(x)$: we have an algorithm which, applied to an object x and a proof that $x \in A$, demonstrates that $P(x)$ holds.

These interpretations are collectively known as the **BHK interpretation**. Note the difference between that interpretation and the classical ones of the connectives and quantifiers. For example, in order to prove the statement $P \vee Q$ classically, it is enough, and often comparatively straightforward, to rule out the possibility that both P and Q are false; in other words, to prove $\neg (\neg P \wedge \neg Q)$. But such an argument generally will not enable us to determine which of the alternatives P and Q actually holds.

Here is a more precise illustration of this point. Define an infinite binary sequence—that is, a sequence $(a_n)_{n \geq 1} \equiv (a_1, a_2, a_3, \ldots)$ whose terms belong to the pair set $\{0, 1\}$—as follows. If $2k + 2$ is a sum of two primes for each positive integer $k \leq n$, set $a_n = 0$; if there exists a positive integer $k \leq n$ such that $2k + 2$ is not a sum of two primes, set $a_n = 1$. Note that, in principle, a finite amount of computation is all that is required to test, for any given positive integer k, whether or not $2k + 2$ can be expressed as a sum of two primes. Thus the sequence $(a_n)_{n \geq 1}$ is computable term-by-term. If we use classical logic, then we easily establish that[1]

$$\forall_n (a_n = 0) \vee \exists_n (a_n = 1),$$

since it is impossible to have both $\neg \forall_n (a_n = 0)$ and $\neg \exists_n (a_n = 1)$. But this classical argument does not enable us to *decide* which of the alternatives $\forall_n (a_n = 0)$ and $\exists_n (a_n = 1)$ holds. In fact, nobody yet knows which of these alternatives holds; a decision on this matter would solve the **Goldbach Conjecture**,

Every even integer ≥ 4 is a sum of two primes,

---

[1] It should be clear what $\forall_n$ and $\exists_n$ mean in such a context.

a statement that has remained unproved, even classically, since it was first discussed in a letter from Goldbach to Euler in 1742.

The use of the Goldbach Conjecture here is purely illustrative. The same argument indicates that the general statement,

> For every binary sequence $(a_n)_{n\geqslant 1}$, either $a_n = 0$ for all n or else there exists n such that $a_n = 1$,

which Bishop dubbed the **limited principle of omniscience** (**LPO**), cannot be proved under the BHK interpretation of the logical connectives and quantifiers. This heuristic justification for regarding **LPO** as an essentially nonconstructive principle—that is, one that cannot be proved under the BHK interpretation— can be made precise: using models of **BISH**, we can show that **LPO** cannot be derived in the intuitionistic predicate calculus; see [88] (Chapter 2, Sections 5 and 6).

One consequence of the exclusion of **LPO** from the canon of constructive principles is that we cannot expect to prove constructively that

$$\forall_{x \in \mathbf{R}} \left(x = 0 \vee x \neq 0 \right), \tag{1.1}$$

where $\mathbf{R}$ denotes the set of real numbers, and $x \neq 0$ signifies that there exists a rational number $r > 0$ such that $|x| > r$.[2] For if $(a_n)_{n\geqslant 1}$ is any binary sequence and we can apply (1.1) to the real number x with binary expansion $0.a_1 a_2 a_3 \ldots$, then either $x = 0$ and therefore $a_n = 0$ for all n, or else there exists $r > 0$ such that $x > r$. In the latter case, choosing a positive integer N such that $2^{-N} < r$ and then testing the terms $a_1, \ldots, a_N$, we are guaranteed to find $n \leqslant N$ such that $a_n = 1$. Hence (1.1) implies **LPO** and so is essentially nonconstructive.

This example clearly has serious implications for the constructive study of $\mathbf{R}$, as it rules the law of trichotomy,

$$\forall_{x \in \mathbf{R}} \left(x < 0 \vee x = 0 \vee x > 0 \right),$$

out of our constructive mathematics. Fortunately, there are constructive substitutes for trichotomy that allow a smooth path through real analysis: working with rational approximations to real numbers, we can justify the **cotransitivity** property of strict inequality,

$$x < y \Rightarrow \forall_{z \in \mathbf{R}} \left(x < z \vee z < y \right).$$

Moreover, defining $\leqslant$ by

$$x \leqslant y \Leftrightarrow \forall_{z \in \mathbf{R}} \left(z < x \Rightarrow z < y \right),$$

---

[2]For the purposes of this book, a real number will be regarded as an object that can be approximated arbitrarily closely by rational numbers.

we have

$$\neg (x > y) \Leftrightarrow x \leqslant y.$$

Note, however, that the proposition

$$\forall_{x,y \in \mathbf{R}} \left(\neg (x \geqslant y) \Rightarrow x < y \right)$$

is equivalent to **Markov's principle (MP)**:

> For each binary sequence $(a_n)_{n \geqslant 1}$, if it is impossible that $a_n = 0$ for all n, then there exists n such that $a_n = 1$.

This principle certainly has dubious computational credentials: why should the impossibility of all terms equalling 0 enable us to compute the index of a term equal to 1? On the other hand, proponents of **MP** might argue that, first, it is almost inconceivable that one could have constructively proved the hypothesis of **MP** without already having proved the conclusion; and, secondly, that if we have proved the hypothesis, then we are guaranteed that a systematic search of the terms $a_1, a_2, \ldots$ will eventually produce a term equal to 1 (even if it takes longer than the remaining life of the universe to do so). We prefer to regard **MP** as essentially nonconstructive, especially as, like **LPO**, it cannot be derived in the intuitionistic predicate calculus.

Another classical principle that, like **LPO**, is recursively false, cannot be derived in the intuitionistic predicate calculus, and is therefore essentially non-constructive is the **lesser limited principle of omniscience (LLPO)**:

> For every binary sequence $(a_n)_{n \geqslant 1}$ which has at most one term equal to 1—that is,
>
> $$\forall_{m,n} \left(m \neq n \Rightarrow a_m a_n = 0 \right)$$
>
> —either $a_{2n} = 0$ for all n or else $a_{2n+1} = 0$ for all n.

Among the consequences of **LLPO** being nonconstructive is that we cannot expect to prove the statement

$$\forall_{x \in \mathbf{R}} (x \geqslant 0 \vee x \leqslant 0) \tag{1.2}$$

constructively: for if we could prove (1.2), then, by applying it to the real number $\sum_{n=1}^{\infty} (-2)^{-n} a_n$ with $(a_n)_{n \geqslant 1}$ any binary sequence with at most one term equal to 1, we could show that either $a_{2n} = 0$ for all n or else $a_{2n+1} = 0$ for all n. Thus even the weak form of trichotomy (1.2) is essentially nonconstructive.

For future reference, we note that the law of excluded middle is constructively equivalent to the statement

$$\neg\neg P \Rightarrow P.$$

Indeed, **LEM** clearly implies the latter. On the other hand, if the latter holds, then since $\neg\neg(P \vee \neg P)$ is easily shown to be a theorem of intuitionistic logic, we obtain **LEM**.

We have shown how certain classically valid statements about real numbers imply **LPO** and are therefore essentially nonconstructive. Other statements are nonconstructive because, sometimes to one's surprise, they imply the full form of the law of excluded middle. For an example of this, we discuss the affirmative constructive notion replacing the classical one of *nonempty set*: we say that a set S is **inhabited** if we can construct an element of S; we then say that S is **inhabited by** that element. Inhabitedness is stronger than the impossibility of being empty. To see this, consider the subset

$$S \equiv \{x : (x = 0 \wedge P) \vee (x = 1 \wedge \neg P)\}$$

of $\{0, 1\}$, where P is any well-defined predicate. We have $\neg (S = \varnothing)$. But if there exists $x \in S$, then, by the BHK interpretation of disjunction, either $x = 0$ and P, or else $x = 1$ and $\neg P$; whence $P \vee \neg P$ holds.

Another example is given by the classical least-upper-bound principle, which states that an inhabited subset S of $\mathbf{R}$ that is bounded above[3] has a **least upper bound**, or **supremum**—that is, a real number σ with the properties

$\triangleright$ σ is an upper bound of S and

$\triangleright$ for each $\sigma' < \sigma$ there exists $s \in S$ with $s > \sigma'$.

(Incidentally, the **greatest lower bound**, or **infimum**, of an inhabited set of real numbers that is bounded below is defined in the obvious analogous way.) To see that this principle entails the law of excluded middle, let P be as before, and consider the set

$$S \equiv \{0\} \cup \{x \in \mathbf{R} : x = 1 \wedge P\}, \tag{1.3}$$

which contains 0 and has 1 as an upper bound. Suppose that the supremum σ of S exists. Although we cannot claim that either $\sigma = 0$ or $\sigma > 0$, using rational numbers to approximate σ within $1/2$, we can show that either $\sigma > 0$ or $\sigma < 1$. In the first case, there exists $x \in S$ such that $x > 0$; whence $x = 1$ and P holds. In the second case we must have $1 \notin S$ and therefore $\neg P$.

An example, like this and others we have given in this section, which shows that under intuitionistic logic a certain classical statement P implies **LEM** or some essentially nonconstructive weak form of **LEM** such as **LPO** or **LLPO**, is known as a **Brouwerian counterexample to** P. It is not a counterexample in the usual sense, for it does not show that P is false (it cannot, since constructive mathematics, in our sense, is consistent with classical mathematics); rather, it

---

[3]We forbear from defining every notion, such as *bounded above*, which should be well known to the reader.

shows that P implies, constructively, some essentially nonconstructive classical principle and so is itself essentially nonconstructive.

Brouwerian counterexamples not only reveal the constructive limitations of classical mathematics, but often suggest ways in which a constructively valid result can be recovered from a classical one that, in its usual form, is essentially nonconstructive. For example, if we examine the foregoing Brouwerian counterexample to the classical least-upper-bound principle, we see that if the set S defined at (1.3) had the property that either $x < 1$ for all $x \in S$ or else there exists $x \in S$ with $x > 0$, then we could compute the supremum σ of S; in the first case, the supremum would be 0, whereas in the second, it would be 1. In fact, this is an illustration of the **constructive least-upper-bound principle**, which we state here for future reference:

> Let S be an inhabited subset of $\mathbf{R}$ that is bounded above. In order for S to have a supremum, it is necessary and sufficient that it be **upper order located** in the sense that for all real numbers a,b with $a < b$, either b is an upper bound of S or else there exists $x \in S$ with $x > a$.

A proof of this theorem, based on a rigorous construction of the real line $\mathbf{R}$, can be found in [29].

1.2   Sets and Functions

Logic alone does not suffice as a foundation for constructive mathematics: we also require concepts of either *type*, or else *set* and *function*. In this book we use a set-theoretic foundation, which we now sketch, leaving out many details whose verification is either routine or follows the classical approach.

It is important to appreciate from the outset that when we collect objects together to form a set, we do so using only objects that have already been constructed; without this restriction, we risk lapsing into impredicativity, where a set is described by means of a property that refers to the set under construction. With our careful restriction to a bottom-up construction of sets at each level using only objects already constructed at a lower level, such freaks as Bertrand Russell's *set of those sets that are not members of themselves* find no place in our theory.

We need some sets to use as the basic building blocks upon which the hierarchy of ever more complicated sets is constructed. Following Errett Bishop [9], we regard the collection $\mathbf{N}^+$ of positive integers $1, 2, 3, \ldots$ as our basic set, and we assume that the positive integers have the usual algebraic and order properties, including mathematical induction.

In general, the construction of a *set* X consists of two parts:

▶ a description of how we construct the elements (members) of X using objects that have been, or could have been, constructed prior to X;

▶ a description of what it means for two elements of X to be equal.

The **equality** on a set X is a binary relation $=$ satisfying the usual properties of an equivalence relation. For the set $\mathbf{Z}$ of all integers, the set $\mathbf{N}$ of all natural numbers $0, 1, 2, \ldots$, and the set $\mathbf{N}^+$, the equality is the relation of identity, in which two elements of the set are regarded as equal if and only if they are identically described. However, the equality on the set $\mathbf{Q}$ of all rational numbers is not that of identity, but is the usual equivalence between rationals:

$$\frac{m}{n} = \frac{m'}{n'} \Leftrightarrow mn' = m'n,$$

where $m, m' \in \mathbf{Z}$ and n, n' are nonzero integers.

We recall that the expression $x \in X$ indicates that x is an element of X; and we use standard notations, such as

$$\{1, 2, 3, \ldots\}$$

to denote $\mathbf{N}^+$, and

$$\{x \in X : P(x)\} \tag{1.4}$$

to denote the set S of those elements x of a pre-existing set X that have the property $P(x)$ (whose description should not involve the set under construction). Taken with the equality relation induced in the obvious manner by that on X, the set (1.4) is regarded as a **subset** of X, the **subset determined by the property** $P(x)$; we also then call X a **superset** of S. We indicate that S is a subset of X by writing $S \subset X$ or $X \supset S$.

Note that we are concerned with properties that are **extensional**, in the sense that for all x, y in S with $x = y$, $P(x)$ holds if and only if $P(y)$ holds. Intensional properties—those that depend on the manner in which objects are presented—are not usually of much interest to us.

In general, more computational information is required to distinguish elements of a set X than to show that elements are equal. In order to do the former, we need an **inequality relation** $\neq$ on X, which is characterised by the properties:

$$x \neq y \Rightarrow \neg\,(x = y),$$
$$x \neq y \Rightarrow y \neq x.$$

We call the inequality

- **nontrivial** if there exist points x, y such that $x \neq y$;

- **tight** if[4]

$$\forall_{x,y \in X}\,(\neg\,(x \neq y) \Rightarrow x = y).$$

---

[4]$\forall_{x,y \in X}$ is a standard shorthand for $\forall_{x \in X}\forall_{y \in X}$.

The standard inequality on $\mathbf{R}$ is given by

$$x \neq y \Leftrightarrow (x > y \vee y > x)$$

and is both nontrivial and tight. Note that $x \neq y$ if and only if $|x - y| > 0$, where the **absolute value** of a real number a is defined as

$$|a| \equiv \max\{a, -a\}.$$

In contrast to the tightness of the inequality on $\mathbf{R}$, the statement

$$\forall_{x \in \mathbf{R}} \left(\neg\, (x = 0) \Rightarrow x \neq 0 \right)$$

implies—indeed, is equivalent to—Markov's principle.

Every set can be equipped with the **denial inequality**, given by $x \neq y$ if and only if $\neg\, (x = y)$. This is seldom of much use constructively, except in the case of a **discrete set** (such as $\mathbf{N}$ or $\mathbf{Q}$)—that is, one satisfying

$$x = y \vee \neg\, (x = y)$$

for all elements x and y.

Two subsets A, B of a set X are **equal sets** if

$$\forall_{x \in X} \left(x \in A \Leftrightarrow x \in B \right).$$

We then write $A = B$. Equality of sets has no meaning for us unless those sets can be provided with comparable equality relations; in other words, unless both the sets being compared are subsets of some larger set.

If $S \subset X$ and $x \in X$, we write $x \notin S$ to signify that it is impossible for x to belong to S. We say that S is **empty** if it cannot be inhabited; we denote the empty subset of any given set X by $\varnothing$. We also write $S \neq \varnothing$ (especially in formulae) to signify that S is inhabited; we emphasise that this is a stronger property than the impossibility of emptiness for S.

A **function** from a set X to a set Y is an algorithm f which produces an element $f(x)$ of Y when applied to an element x of X, and which is extensional: that is, $f(x) = f(x')$ in Y whenever $x = x'$ in X. A function from X to Y is also called a **mapping**, or **map**, of X into Y. The notation

$$f : X \to Y$$

indicates that f is a mapping of X into Y. The set X is called the **domain** of f. When the domain of the function f is clearly understood, we often denote that function by

$$x \rightsquigarrow f(x)$$

or by

$$f : x \rightsquigarrow f(x).$$

If S is a subset of X, then the **restriction of f to S** is the mapping $f|_S :$ $x \rightsquigarrow f(x)$ taken with domain S. If g is a mapping of S into Y, then a mapping $f : X \to Y$ is an **extension of g to X** if g is the restriction of f to S.

The set
$$f(X) \equiv \{y \in Y : \exists_{x \in X} (y = f(x))\}$$
is called the **range** of f. If the range of f equals the whole set Y, we say that f **maps X onto Y**.

We denote by Y^X the set of all functions from X into Y. If Y has an inequality relation, then Y^X has a corresponding inequality defined by
$$\forall_{f,g \in Y^X} (f \neq g \Leftrightarrow \exists_{x \in X} (f(x) \neq g(x))) .$$

If X also has an inequality relation, then an important property applicable to an element f of Y^X is that of **strong extensionality**:
$$\forall_{x,x' \in X} (f(x) \neq f(x') \Rightarrow x \neq x') .$$

The strong extensionality of all mappings from $\mathbf{R}$ to $\mathbf{R}$ implies Markov's principle.

A mapping $f : X \to Y$ is called **one-one** if $x = x'$ whenever $f(x) = f(x')$. Note that when X, Y have inequality relations, we say that f is **injective** if $f(x) \neq f(x')$ in Y whenever $x \neq x'$ in X; if the inequality on X is tight, f is then one-one.

The **composition**, or **composite**, of two mappings $f : X \to Y$ and $g : Y \to Z$ is the mapping $g \circ f : X \to Z$ defined by
$$g \circ f(x) \equiv g(f(x))$$
for all $x \in X$.

The supremum and infimum (when they exist) of a bounded subset S of $\mathbf{R}$ are denoted, respectively, by $\sup S$ and $\inf S$. A mapping f of a set X into $\mathbf{R}$ is **bounded** if $f(X)$ is a bounded subset of $\mathbf{R}$; in which case, the supremum of $f(X)$ is called the **supremum** of the function f and is denoted by $\sup f$; a similar definition applies for the **infimum** of f, denoted by $\inf f$.

The **Cartesian product** $X_1 \times X_2$ of two sets, each with an inequality relation, is constructed as follows. The elements of $X_1 \times X_2$ are ordered pairs of the form (x_1, x_2) with $x_1 \in X_1$ and $x_2 \in X_2$. The equality and inequality on $X_1 \times X_2$ are defined respectively by
$$(x_1, x_2) = (y_1, y_2) \Leftrightarrow (x_1 = y_1 \wedge x_2 = y_2),$$
$$(x_1, x_2) \neq (y_1, y_2) \Leftrightarrow (x_1 \neq y_1 \vee x_2 \neq y_2) .$$

We refer colloquially to X_1 and X_2 as "factors" of X. These definitions readily extend to cover the Cartesian product $X_1 \times \cdots \times X_n$ of finitely many sets $X_1, \ldots, X_n$. We adopt the convention that, for example, $\mathbf{x}$ denotes the element of $X_1 \times \cdots \times X_n$ whose ith component is x_i, and $\mathbf{y}$ the element whose ith component is y_i. The kth **projection** of the Cartesian product is the mapping $\mathrm{pr}_k : \mathbf{x} \rightsquigarrow x_k$ of $X_1 \times \cdots \times X_n$ onto X_k.

The Cartesian product $X \times X$ of a set X with itself is also written X^2. A subset of X^2 is called a **binary relation on** X. More generally, a subset of a Cartesian product $X_1 \times X_2$ is called a **relation between elements of** X_1 **and elements of** X_2. For example, we shall be concerned in Chapter 2 with a relation $\bowtie$ between points of a set X and subsets of X.

A **family** of elements of a set X is a mapping $n \rightsquigarrow x_n$ from a set $\mathfrak{D}$—the **index set** of the family—into X; we use the notation $(x_n)_{n \in \mathfrak{D}}$ for such a family. When $\mathfrak{D} = \mathbf{N}^+$ (respectively, $\mathbf{N}$), the family is called a **sequence** in X, with nth **term** x_n, and is also written $(x_n)_{n \geqslant 1}$ (respectively, $(x_n)_{n \geqslant 0}$). A **subsequence** of a sequence $(x_n)_{n \geqslant 1}$ consists of $(x_n)_{n \geqslant 1}$ itself and a sequence $(n_k)_{k \geqslant 1}$ of natural numbers such that $n_1 < n_2 < \cdots$; we identify this subsequence with the sequence $(x_{n_k})_{k \geqslant 1}$ whose kth term is x_{n_k}.

A **finite sequence of length** N, where $N \in \mathbf{N}^+$, is a mapping x with domain $\{1, 2, \ldots, N\}$. This mapping can be identified with the **ordered** N-**tuple** $(x_1, x_2, \ldots, x_N)$ where $x_i = x(i)$.

Let $(S_n)_{n \in \mathfrak{D}}$ be a family of subsets of a set X. We define its **union** to be

$$\bigcup_{n \in \mathfrak{D}} S_n \equiv \{x \in X : \exists_{n \in \mathfrak{D}} \, (x \in S_n)\},$$

and its **intersection** to be

$$\bigcap_{n \in \mathfrak{D}} S_n \equiv \{x \in X : \forall_{n \in \mathfrak{D}} \, (x \in S_n)\}.$$

These are also denoted by

$$\bigcup \{S_n : n \in \mathfrak{D}\}$$

and

$$\bigcap \{S_n : n \in \mathfrak{D}\},$$

respectively. When the family is a finite sequence of length N, we also write $\bigcup_{n=1}^{N} S_n$ or $S_1 \cup \cdots \cup S_N$ to denote its union, and $\bigcap_{n=1}^{N} S_n$ or $S_1 \cap \cdots \cap S_N$ its intersection. Two subsets of X are said to be **disjoint** if their intersection is empty. When the family is an infinite sequence $(S_n)_{n \geqslant 1}$, we denote the union and intersection respectively by $\bigcup_{n \geqslant 1} S_n$ and $\bigcap_{n \geqslant 1} S_n$.

Unions and intersections have many, but not all, of the properties familiar from classical mathematics. We shall point out significant differences between their classical and constructive properties as they arise below.

We need a few facts about the cardinality of sets. A set S is **finitely enumerable** if there exist a nonnegative integer n and a mapping f from $\{k \in \mathbf{N}^+ : k \leqslant n\}$ onto S; if this map is also one-one, we say that S is **finite**. In the first case we also say that S has **at most** n **elements,** and in the second that S has **(exactly)** n **elements.** Note that the empty set is finitely enumerable.

Although *finite* implies (and is classically equivalent to) *finitely enumerable*, the converse does not hold constructively, since the statement

Every finitely enumerable subset of $\mathbf{R}$ is finite

entails **LPO**: to see this, given a binary sequence $(a_n)_{n \geqslant 1}$, consider the finitely enumerable set

$$\left\{ 0, \sum_{n=1}^{\infty} 2^{-n} a_n \right\}.$$

A set S is **countable** if there exists a mapping from $\mathbf{N}^+$ onto S. Inhabited finitely enumerable sets are countable, as is the Cartesian product of two countable sets. A subset S of a set X with an inequality relation is **infinite** if for each finite subset F of S, there exists $s \in S$ such that $s \neq x$ for each $x \in F$; note that this is a stronger constructive property than the denial of finiteness.

The classical **axiom of choice** says that if S is a subset of $A \times B$ and for each $x \in A$ there exists $y \in B$ such that $(x, y) \in S$, then there is a function f from A to B such that $(x, f(x)) \in S$ for each $x \in A$. It is often thought that rejection of the axiom of choice characterises constructive mathematics. This is not so: constructive mathematics is characterised *positively*, by the BHK interpretation and hence by intuitionistic logic. Nevertheless, as was shown by Diaconescu [38] and Goodman and Myhill [45], the axiom of choice implies the law of excluded middle.

Following the custom of many, if not most, workers in constructive mathematics, we freely use the **principle of dependent choice**:

If $a \in A$ and $S \subset A \times A$, and if for each $x \in A$ there exists $y \in A$ such that $(x, y) \in S$, then there exists a sequence $(a_n)_{n \geqslant 1}$ in A such that $a_1 = a$ and $(a_n, a_{n+1}) \in S$ for each n.

A consequence of this principle is the **principle of countable choice**: the case $A = \mathbf{N}^+$ of the full axiom of choice.

We should point out that some authors, notably Richman, have expressed serious doubts about the constructive validity of dependent choice, and have shown how to develop parts of constructive analysis without even the principle of countable choice [77].

1.3   Quasi-metric and Topological Spaces

Although we assume familiarity with the classical theories of metric, normed, and topological spaces, it would be unwise of us not to provide the fundamental definitions, as well as one or two distinctive aspects, of the counterpart constructive theories. We omit proofs here, referring the reader to such sources as [9, 12, 24, 29, 88] for the details.

A **quasi-metric space** is a pair (X, ρ) consisting of an inhabited set X and a mapping $\rho : X \times X \to \mathbf{R}$ such that for all x, y, z in X,

M1 $\rho(x,y) = 0$ if and only if $x = y$;

M2 $\rho(x,y) > 0 \Rightarrow \rho(y,x) > 0$;

M3 $\rho(x,z) \leqslant \rho(x,y) + \rho(y,z)$ (the **triangle inequality**).

The mapping ρ is called a **quasi-metric** on X.

It follows from these axioms that $\rho(x, y) \geqslant 0$ for all x and y: suppose that $\rho(x,y) < 0$; then, by **M2**, $\rho(y,x) \leqslant 0$, so, by **M1** and **M3**,

$$0 = \rho(x, x) \leqslant \rho(x, y) + \rho(y, x) < 0,$$

a contradiction from which it follows that $\rho(x,y) \geqslant 0$. If also $\rho(x, y) = \rho(y,x)$ for all x and y, then ρ is called a **metric**, and (X, ρ)—or, if it is clear which metric we are dealing with, just X itself—a **metric space**. We define the inequality $\neq$ on a quasi-metric space (X, ρ) by

$$x \neq y \Leftrightarrow \rho(x, y) > 0.$$

This inequality is tight: for if $\neg (x \neq y)$, then $\neg (\rho(x, y) > 0)$ and therefore $\rho(x,y) \leqslant 0$; whence $\rho(x,y) = 0$ and so $x = y$.

For $n \geqslant 1$ the set $\mathbf{R}^n$ is a metric space relative to the **Euclidean metric**

$$\rho\left(\mathbf{x}, \mathbf{y}\right) \equiv \sqrt{\sum_{k=1}^{n} |x_k - y_k|^2}.$$

(Recall here our notation for elements of a Cartesian product.)

A subset S of X, taken with the restriction of ρ to $S \times S$, is a quasi-metric space, and is called a (quasi-metric) **subspace** of X.

Let (X_k, ρ_k) $(1 \leqslant k \leqslant n)$ be quasi-metric spaces. Then their Cartesian product $X \equiv X_1 \times \cdots \times X_n$ carries a natural **product quasi-metric** defined by

$$(\mathbf{x}, \mathbf{y}) \rightsquigarrow \max \left\{ \rho(x_k, y_k) : 1 \leqslant k \leqslant n \right\}.$$

Equipped with this quasi-metric, X becomes the **product of the quasi-metric spaces** $X_1, \ldots, X_n$.

For each element a of a quasi-metric space X and for each $r > 0$, the **open ball with centre a and radius r** in X is

$$B(a, r) \equiv \{x \in X : \rho(a, x) < r\},$$

and the **closed ball with centre a and radius r** in X is

$$\overline{B}(a, r) \equiv \{x \in X : \rho(a, x) \leqslant r\}.$$

The **interior** of a subset S of X is defined as

$$S^\circ \equiv \{x \in X : \exists_{r>0} \, (B(x,r) \subset S)\},$$

and the **closure** of S by

$$\overline{S} \equiv \{x \in X : \forall_{\varepsilon>0} \exists_{s\in S} \, (\rho(x,s) < \varepsilon)\}.$$

We say that S is **open** (respectively, **closed**) in X if $S = S^\circ$ (respectively, $S = \overline{S}$). Note that in the metric space $\mathbf{R}$ (with the standard, Euclidean metric), although the complement of an open set is closed, a Brouwerian counterexample—found, together with several others about the metric structure on $\mathbf{R}$, in [15]—shows that we cannot prove that the complement of a closed set is open.

A sequence $(x_n)_{n \geqslant 1}$ in the quasi-metric space X is said to be **convergent** if there exists a (perforce unique) element x of X, called the **limit** of the sequence, such that

$$\forall_{\varepsilon>0} \exists_N \forall_{n \geqslant N} \, (\rho(x,x_n) < \varepsilon).$$

We then write

$$x_n \to x \text{ as } n \to \infty$$

or

$$\lim_{n \to \infty} x_n = x.$$

On the other hand, we say that $(x_n)_{n \geqslant 1}$ is a **Cauchy sequence** in X if

$$\forall_{\varepsilon>0} \exists_N \forall_{m,n \geqslant N} \, (\rho(x_m,x_n) < \varepsilon).$$

Every convergent sequence is a Cauchy sequence. If every Cauchy sequence in X converges to a limit in X, then we say that X is a **complete** metric space. Contrary to a common belief, the completeness of $\mathbf{R}$ with respect to the standard metric holds constructively. As classically, a complete subset of a metric space is closed in that space, and a closed subset of a complete metric space is complete.

Given $\varepsilon > 0$, by an ε-**approximation** to a metric space X we mean an inhabited subset T of X such that

$$\forall_{x\in X} \exists_{t\in T} \, (\rho(x,t) < \varepsilon).$$

If for each $\varepsilon > 0$ there exists a finitely enumerable ε-approximation to X, then we say that X is **totally bounded**; if also X is complete, then we call it a **compact (metric) space**. If S is an inhabited, totally bounded subset of $\mathbf{R}$, then $\sup S$ and $\inf S$ exist (see Proposition 3.3.6 below).

A mapping $f : X \to Y$ between quasi-metric spaces is **uniformly continuous** if

$$\forall_{\varepsilon>0} \exists_{\delta>0} \forall_{x,y\in X} \, (\rho(x,y) < \delta \Rightarrow \rho(f(x),f(y)) < \varepsilon).$$

In that case, if X is a totally bounded metric space, then so is $f(X)$.

A metric space X is called **locally totally bounded** if for each bounded subset S of X there exists a totally bounded set $T \subset X$ with $S \subset T$. In that case, for each $x \in X$, the balls $B(x,r)$ and $\overline{B}(x,r)$ are totally bounded for all but countably many $r > 0$. If X is both locally totally bounded and complete, then we call it **locally compact**.

A subset S of a metric space (X,ρ) is said to be **located** in X if the distance

$$\rho(x, S) \equiv \inf \{\rho(x, s) : s \in S\}$$

exists for each $x \in X$. By considering the set S defined at (1.3) we can show that if every inhabited subset of $\{0,1\}$ is located, then **LEM** is derivable. However, every locally totally bounded subset of a metric space is located; conversely, every located subset of a locally totally bounded space is locally totally bounded.

We now provide a fundamental example of a metric space. Let X be a vector space over the groundfield $\mathbf{K}$ (which stands for either $\mathbf{R}$ or $\mathbf{C}$). Following [29], we require X to be equipped with a tight inequality $\neq$ that is **compatible** with the algebraic structure, in the sense that for all $x, y \in X$ and all $\lambda \in \mathbf{K}$,

$$
\begin{aligned}
x \neq y &\Leftrightarrow x - y \neq 0, \\
x + y \neq 0 &\Rightarrow x \neq 0 \vee y \neq 0, \text{ and} \\
\lambda x \neq 0 &\Rightarrow \lambda \neq 0 \wedge x \neq 0.
\end{aligned}
$$

It readily follows from the first of these properties that

$$x \neq y \Rightarrow \forall_{z \in X} (x + z \neq y + z).$$

The requirement of compatibility between the inequality and the linear structure is a natural one and is automatically fulfilled by the denial inequality under classical logic. In the case $X = \mathbf{K}$ the standard inequality is compatible with the linear structure.

By a **seminorm** on X we mean a nonnegative mapping $\| \ \| : x \rightsquigarrow \|x\|$ of X into $\mathbf{R}$ such that for all x,y in X and all λ in $\mathbf{K}$,

- $\|x\| > 0 \Rightarrow x \neq 0$,

- $\|\lambda x\| = |\lambda| \, \|x\|$, and

- $\|x + y\| \leqslant \|x\| + \|y\|$.

We call the pair $(X, \| \ \|)$—or, when no confusion is likely, just X itself—a **seminormed (linear) space** over $\mathbf{K}$. If $x \in X$ and $\|x\| > 0$, then x is called a **nonzero vector**. We call the seminormed space X **nontrivial** if it contains a nonzero vector.

If the inequality on the seminormed space X satisfies

$$\forall_{x \in X} (x \neq 0 \Leftrightarrow \|x\| > 0), \tag{1.5}$$

then we call $\| \ \|$ a **norm** on X, $\|x\|$ the **norm of the vector** x, and $(X, \| \ \|)$—or just X—a **normed (linear) space** over $\mathbf{K}$. Note that every seminorm $\| \ \|$ on a linear space X induces an inequality relation—namely, the one defined by (1.5)—with respect to which $\| \ \|$ becomes a norm.

It remains in this chapter to define some elementary general topological notions that are doubtless familiar to the reader. By a **topology** on a set X we mean a set τ of subsets of X with the following properties:

T1 $X \in \tau$ and $\varnothing \in \tau$.

T2 The union of any family of sets in τ belongs to τ.

T3 The intersection of two (and hence, by induction, any finite number of) elements of τ belongs to τ.

The pair (X, τ)—or, when there is no doubt about the topology under discussion, the set X itself—is called a **topological space**, and the elements of τ are then called the **open subsets** of X. A **base** for the topology τ is a family $\mathfrak{B}$ of open sets such that every inhabited subset of X is open if and only if it is a union of sets in $\mathfrak{B}$. A family S of subsets of X is called a **subbase** for τ if the finite intersections of members of S form a base for τ.

The **interior** S° of a subset S of X consists of all points $x \in X$ for which there exists $U \in \tau$ such that $x \in U \subset S$. The **neighbourhoods** of a point x in X are those sets S whose interior contains x. A **neighbourhood base**, or **base of neighbourhoods**, for $x \in X$ is a family $\mathfrak{N}$ of neighbourhoods of x such that for each neighbourhood A of x there exists $U \in \mathfrak{N}$ with $x \in U \subset A$. The standard (quasi-metric) topology on a quasi-metric space (X, ρ) is the topology in which for each point x, the open balls with centre x form a neighbourhood base.

The **closure** $\overline{S}$ of S consists of all points $x \in X$ such that every neighbourhood of x intersects S; and (as with a metric space) S is said to be **closed** if it equals its closure. A set D is **dense** in X if $X = \overline{D}$. If X has a countable dense set, then X is said to be **separable**. A located subset of a separable metric space is separable.

A mapping $f : X \to Y$ between topological spaces is **(topologically) continuous** if $f^{-1}(V)$ is open in X for each open $V \subset Y$.

Each subset Y of X has a corresponding **subspace topology** τ_Y consisting of all sets of the form $Y \cap U$ with U open in X. The topological space (Y, τ_Y) is then called a (topological) **subspace** of X.

If X_1 and X_2 are topological spaces, then the sets of the form $A_1 \times A_2$, with A_k open in the space X_k, form a base for a topology—called the **product topology**—on their Cartesian product. Taken with the product topology, $X_1 \times X_2$ is known as the (topological) **product** of X_1 and X_2.

Finally, let τ, τ' be topologies on X. We say that τ is **finer** than τ', and that τ' is **coarser** than τ, if $\tau' \subset \tau$.

Although topology plays a significant part in the succeeding chapters of our book, we shall be more interested in the notions of apartness—between points and points (that is, inequality), points and sets, or sets and sets—and of uniformity.

Notes on Chapter 1

The history of the theory of proximity up to 1970 is outlined on pages 1–6 of [75]; for more recent history, see [7]. The former book is a good source of information about the theory itself; the forthcoming monograph [74] will doubtless be an even better one.

Why do we describe the notion of apartness as "computationally more informative than that of proximity"? Consider what it means, constructively and informally, for two real numbers x and y to be distinct: $x \neq y$ if and only if we can compute a rational number r that we can place strictly between x and y. On the other hand, x and y are equal as real numbers if and only if there is no such rational number r. In the same way, to prove that two subsets of $\mathbf{R}$ are apart, we need to compute a positive lower bound for the distance between points of one set and points of the other; whereas to prove the proximity of those sets, we need to show that the lower bounds for such distances are never positive. Thus apartness deals with the existence of certain objects (in the examples here, real numbers), but proximity deals with the non-existence of such objects.

The name *BHK interpretation* comes from Brouwer–Heyting–Kolmogorov, the names of the three mathematicians who independently, and in Brouwer's case perhaps implicitly, came up with that interpretation; see [32, 33, 54, 63]. The axioms for intuitionistic logic, published by Heyting in 1930 [53], were intended to capture that interpretation formally.

The BHK interpretation of implication, while more natural than the classical one of *material implication* in which $(P \Rightarrow Q)$ is equivalent to $(\neg P \vee Q)$, has not satisfied all researchers using constructive logic. Shortly before he died, Errett Bishop communicated to Douglas Bridges his discontent with the standard constructive interpretation of implication. Unfortunately, he left nothing more than very rudimentary sketches of his ideas for its improvement.

In this book, our interpretation of the phrase *constructive mathematics* is that of Bishop [9]: it is mathematics with intuitionistic logic and an appropriate set theory such as those found in [2, 3, 43, 72]. This type of constructive mathematics, which we have called **BISH**, has three standard models:[5]

Brouwer's intuitionistic mathematics (**INT**), which we can regard as **BISH** plus certain principles introduced by Brouwer [40];

---

[5]We use the word *model* here in a highly informal sense.

The recursive constructive mathematics (**RUSS**) of the Russian school founded by Markov; this is essentially **BISH** plus the Church–Markov–Turing thesis (see Chapter 3 of [24]) and Markov's principle [64];

Classical mathematics (**CLASS**), which is **BISH** plus the law of excluded middle (and therefore full classical logic).

For more information about the first two of these models, see [6, 24, 88]. Another model is Weihrauch's Type II Effectivity Theory of "computable analysis"; Bauer [5] has shown how to translate results from **BISH** into that framework.

Since **LPO** holds in **CLASS** but is provably false in **INT** and **RUSS**, it cannot be derived in **BISH**. A Kripke model shows that Markov's principle, and therefore **LPO**, cannot be derived in the intuitionistic predicate calculus; see Chapter 5 of [40], or pages 137–138 of [24].

Bishop used the term *subfinite* where we use *finitely enumerable*.

Locatedness plays a vital role in the constructive theory of metric and normed spaces. For example, a nonzero bounded linear functional u on a normed space X has a norm

$$\|u\| \equiv \sup\left\{|u(x)| : x \in X, \|x\| \leqslant 1\right\}$$

if and only if its kernel

$$\ker(u) \equiv \{x \in X : u(x) = 0\}$$

is located; and the Hahn–Banach extension theorem depends on the locatedness of the kernel of the function to be extended. In Chapter 3, we discuss how the notion of locatedness can be lifted from the context of a metric space to more abstract ones. The absence of a distance function in the latter context makes this lifting nontrivial within constructive mathematics.

Compactness is a problematic notion in **BISH**. The problems with compactness are that (i) the open-cover definition does not apply to the interval $[0,1]$ in the recursive model and (ii) sequential, or filter, compactness is completely nonconstructive; so the only classical notion that applies to a quasi-uniform space (see Section 3.2 below) is that of total boundedness and completeness, which cannot obviously be generalised to apply outside that context.

2

POINT-SET APARTNESS

... what person can dare say they cannot afford to take time for apartness —indeed, who can afford not to take time for apartness?

M. Basil Pennington

Synopsis

We first introduce the notion of a (pre-)apartness between points and subsets in an abstract space X, and derive some elementary properties from our axioms. Each point-set pre-apartness gives rise to a topology—the apartness topology—on X, and to several constructively distinct continuity properties, which are explored in Section 2.3. Limits and the Hausdorff property are discussed in Section 2.4, and product pre-apartness spaces in Section 2.5. In the final section we discuss the role of impredicativity in our theory.

2.1   Pre-apartness

Throughout this chapter our basic structure will be an inhabited set X equipped with an inequality relation $\neq$, which in this context can also be called a *point-point apartness*. A subset S of X has two natural complementary subsets:

▷ the **logical complement**

$$\neg S \equiv \{x \in X : \forall_{y \in S} \neg (x = y)\},$$

▷ and the **complement**

$$\sim S \equiv \{x \in X : \forall_{y \in S} \, (x \neq y)\}.$$

The properties of $\neq$ ensure that $\sim S \subset \neg S$. For each subset A of X we write

$$A \sim S \equiv A \cap \sim S.$$

For the record we note that although $S \subset \neg\neg S$ and $\neg S \cup \neg T \subset \neg(S \cap T)$, in neither case can we prove the reverse inclusions constructively; the same applies with $\neg$ replaced by $\sim$.

We require our inhabited set X to carry also a special relation, $\bowtie$, between points x and subsets S. If $x \bowtie S$, we say that x is **apart from** S. We shall write down the axioms for $\bowtie$ in a moment; but for convenience we introduce here the **apartness complement**

$$-S \equiv \{x \in X : x \bowtie S\}$$

of S, and the notation
$$A - S \equiv A \cap -S.$$

As we shall see, in a quasi-metric space (X, ρ) the apartness complement $-S$ is the set of those $x \in X$ that are bounded away from S.

The following are the axiomatic properties, holding for all $x \in X$ and all subsets A, B of X, that we require of a **pre-apartness** $\bowtie$:

A1 $x \bowtie \varnothing$

A2 $-A \subset \sim A$

A3 $x \bowtie (A \cup B) \Leftrightarrow x \bowtie A \wedge x \bowtie B$

A4 $-A \subset \sim B \Rightarrow -A \subset -B.$

We then call the pair $(X, \bowtie)$ —or, when the apartness relation is clearly understood, just the set X itself—a **pre-apartness space**, and the data defining the relations $\neq$ and $\bowtie$ the **pre-apartness structure** on X. When we want to emphasise that the pre-apartness is associated with a particular set X, we sometimes denote[1] it by $\bowtie_X$.

Axiom **A1** says that $X = -\varnothing$. Since $-X \subset \sim X$ (by axiom **A2**) and therefore $-X \subset \neg X = \varnothing$, we see that $\varnothing = -X$. As a special case of axiom **A2**, we have

$$\forall_{x,y \in X} \, (x \bowtie \{y\} \Rightarrow x \neq y).$$

---

[1]The reader is warned that we apply subscripts with different meanings to an apartness. For example, if τ is a topology, then we denote an important related pre-apartness relation by $\bowtie_\tau$. In practice, it should be clear whether a subscript attached to the symbol $\bowtie$ signifies a set, a topology, or—in Chapter 3—a quasi-uniform structure.

Also, **A3** is equivalent to

$$-(A \cup B) = -A \cap -B.$$

Axiom **A4** may look a little mysterious, so we pause to show where it comes from. In the classical theory of proximity spaces [35], the primitive point-set notion is a binary relation δ of *proximity*, and what we call *apartness* is the negation of proximity. One of the conditions that can be imposed on a proximity is the **Lodato property**,

$$(x \,\delta\, B \wedge \forall_{y \in B}(y \,\delta\, A)) \Rightarrow x \,\delta\, A,$$

where, for example, $x \,\delta\, A$ is shorthand for $\{x\} \,\delta\, A$. Under classical logic, since $\bowtie$ is the denial of δ, this last implication is equivalent to

$$x \bowtie A \Rightarrow (x \bowtie B \vee \exists_{y \in B}(y \bowtie A))$$

and therefore to

$$(x \bowtie A \wedge \forall_y (y \bowtie A \Rightarrow y \notin B)) \Rightarrow x \bowtie B.$$

In the classical theory of nearness, where $\sim$ and $\neg$ would coincide, this last statement is just **A4** in disguise.

If, in addition to **A1–A4**, the pre-apartness satisfies

A5 $x \bowtie A \Rightarrow \forall_{y \in X} (x \neq y \vee y \bowtie A),$

then we call it an **apartness**, and the space X an **apartness space**. The value of the additional axiom **A5** is that it provides us with (classically automatic) alternatives that facilitate many constructive proofs. Nevertheless, it makes good sense to work without **A5** where it is not needed.

The canonical example of a full-blooded apartness space is a quasi-metric space (X, ρ) with the pre-apartness defined by

$$x \bowtie A \Leftrightarrow \exists_{r>0} \forall_{y \in A} (\rho(x, y) \geqslant r). \tag{2.1}$$

In this case, the apartness complement of a subset A of X,

$$-A \equiv \{x \in X : \exists_{r>0} \forall_{y \in A} (\rho(x, y) \geqslant r)\},$$

is also known as the **metric complement** of A. It is routine to verify that $\bowtie$ has properties **A1–A3**. For **A4**, let $x \bowtie A$ and $-A \subset {\sim}B$. Choose $r > 0$ such that $\rho(x, z) \geqslant r$ for all $z \in A$. If $y \in B$ and $\rho(x, y) < r/2$, then for all $z \in A$ we have

$$\rho(y, z) \geqslant \rho(x, z) - \rho(x, y) > r/2, \tag{2.2}$$

so $y \in -A$ and therefore $y \in {\sim}B$, which is absurd. Hence $\rho(x, y) \geqslant r/2$ for all $y \in B$, and therefore $x \bowtie B$. This completes the verification of **A4**. For

A5 we let x, A, and r be as before. Then for each $y \in X$, either $\rho(x,y) > 0$ and therefore $x \neq y$, or else $\rho(x,y) < r/2$. In the second case, for all $z \in A$ we have (2.2) and therefore $y \bowtie A$.

The apartness defined at (2.1) is called the **quasi-metric apartness** on X, and X, taken with this apartness, is called a **quasi-metric apartness space**. When ρ is actually a metric, we replace *quasi-metric* by *metric* in both places in the preceding sentence.

Classically, with the law of excluded middle at hand, it is easy to prove that every pre-apartness space satisfies **A5**. Is this the case constructively? First note that if X is any set with an inequality relation, then

$$\forall_{x \in X} \forall_{A \subset X} (x \bowtie A \Leftrightarrow x \in \sim A)$$

defines a pre-apartness on X. Now take $X = [0,1]$ with the usual inequality relation and the pre-apartness just defined, and take $A = (0,1]$. Then $0 \bowtie A$. Given a binary sequence $(a_n)_{n \geqslant 1}$ with at most one term equal to 1, define

$$y \equiv \sum_{n=1}^{\infty} \frac{a_n}{n} \in X.$$

If $y \neq 0$, then $a_n = 1$ for some n; if $y \bowtie A$, then $a_n = 0$ for all n. Thus if axiom **A5** holds in the pre-apartness space X, we can derive **LPO**.

We conclude from this example that **A5** cannot be derived constructively from **A1–A4**. On the other hand, since **A5** holds in a metric apartness space, its negation cannot be derived from **A1–A4**. Thus **A5** is constructively independent of **A1–A4**.

In the remaining results of this section, except in some of the examples, we assume that X is a pre-apartness space, that $x,y,z,\ldots$ are points of X, and that $A,B,\ldots$ are subsets of X. If we want to specialise to an apartness space, we shall make it explicit that we are also assuming **A5**.

Proposition 2.1.1 *If $A \subset B$, then $-B \subset -A$.*

Proof. Since $A \cup B = B$, the conclusion follows from **A3**. ∎

Proposition 2.1.2 $A \subset \sim - A$.

Proof. By **A2**, $-A \subset \sim A$; whence $A \subset \sim\sim A \subset \sim - A$. ∎

Proposition 2.1.3 *If $\sim A \subset \sim B$, then $-A \subset -B$.*

Proof. Since $-A \subset \sim A$, this is a simple application of **A4**. ∎

If Y is an inhabited subset of X, we have a natural inequality $\neq_Y$ and point-set relation $\bowtie_Y$ defined for $y,y' \in Y$ and $S \subset Y$ by

$$y \neq_Y y' \Leftrightarrow y \neq y',$$
$$y \bowtie_Y S \Leftrightarrow y \bowtie S, \tag{2.3}$$

where on the right side, $\neq$ and $\bowtie$ are the original inequality and pre-apartness on X. We say that $\neq_Y$ and $\bowtie_Y$ are **induced** on Y by their counterparts on X. It is easy to show that $\bowtie_Y$ satisfies **A1–A3**. If, in addition, it satisfies **A4** in the form

$$(Y - A \subset Y \sim B) \Rightarrow (Y - A \subset Y - B),$$

then it is a pre-apartness on Y; in that case, taken with the induced inequality and pre-apartness, Y is called a **pre-apartness subspace** of X; if $\bowtie_Y$ satisfies **A5**, then we call Y an **apartness subspace** of X. We usually omit the subscript, and denote the induced inequality and apartness on Y simply by $\neq$ and $\bowtie$.

Proposition 2.1.4 *Let $(X, \bowtie)$ be a pre-apartness space satisfying the condition*

$$\forall_{x,y \in X} \forall_{S \subset X} ((x \in -S \wedge y \notin -S) \Rightarrow x \neq y). \tag{2.4}$$

Then every inhabited subset of X is a pre-apartness subspace.

Proof. Let Y be an inhabited subset of X, and let A,B be subsets of Y such that $Y - A \subset Y \sim B$. If $y \in B \subset Y$, then $y \notin X - A$, so, by (2.4), $y \neq x$ for each $x \in X - A$. Thus $X - A \subset X \sim B$. By **A4** in the space X, we have $X - A \subset X - B$; from which it follows that $Y - A \subset Y - B$. ∎

We refer to condition (2.4) as the **reverse Kolmogorov property**. It is equivalent to the condition

$$\forall_{S \subset X} (\neg - S = \sim - S). \tag{2.5}$$

If X is an apartness space and Y is an inhabited subset of X, then, since **A5** both implies condition (2.4) and is inherited from $\bowtie_X$ by $\bowtie_Y$, it follows, with reference to Proposition 2.1.4, that Y is also an apartness space.

The following results about double and triple complements may appear somewhat unattractive to the reader unused to working constructively, but are occasionally useful.

Although any subset A of a set with an inequality is contained in its double complement $\sim\sim A$, constructively the double complement looks potentially larger than the set A itself, and we cannot in general prove that these two sets are equal. However, it is an easy exercise to prove that

$$\sim A = \sim(\sim\sim A) = \sim\sim(\sim A) = \sim\sim\sim A.$$

If A is a subset of our pre-apartness space X we can say more.

Proposition 2.1.5 $-A = -\sim\sim A = -\sim- A.$

Proof. Since $-A \subset \sim\sim - A$, we see from **A4** that

$$-A \subset -\sim- A. \tag{2.6}$$

Also, by Proposition 2.1.1,

$$-\sim\sim A \subset -A. \tag{2.7}$$

On the other hand, since $-A \subset \sim A$, we have $\sim\sim A \subset \sim - A$ and therefore (again by Proposition 2.1.1)

$$-\sim - A \subset -\sim\sim A. \tag{2.8}$$

Putting together (2.6)–(2.8), we obtain the desired identities. ∎

We say that the point $x \in X$ is **near** the set $A \subset X$, and we write near(x, A), if

$$\forall_{S \subset X} \, (x \in -S \Rightarrow \exists_{y \in X} \, (y \in A - S)).$$

Although our focus throughout the book is primarily on apartness, nearness will arise from time to time—notably, to characterise the closure of a set relative to the pre-apartness, and in our discussions of limits and product apartness spaces.

Proposition 2.1.6 *If near(x, A) in X, then A is inhabited.*

Proof. By **A1**, $x \in -\varnothing$. So if near (x, A), then, by the definition of *near*, there exists y in $A - \varnothing$, which equals A. ∎

Proposition 2.1.7 *In the case where X is a metric apartness space, near (x, A) if and only if $A \cap B(x,r)$ is inhabited for each $r > 0$.*

Proof. Suppose that near (x, A). Given $r > 0$, let

$$S \equiv \{y \in X : \rho(x, y) \geqslant r/2\} .$$

Then

$$x \in -S \subset \overline{B}(x, r/2) \subset B(x, r).$$

Hence, by definition of the nearness predicate, there exists $y \in A - S \subset A \cap B(x,r)$.

Now suppose, conversely, that $A \cap B(x,r)$ is inhabited for each $r > 0$, and that $x \in -S$. Then there exists $r > 0$ such that $\rho(x,s) \geqslant r$ for each $s \in S$; whence $B(x,r) \subset -S$ and therefore $A - S$ is inhabited. Since S is arbitrary, we conclude that near (x, A). ∎

Proposition 2.1.8 *If $x \in A$, then near(x, A).*

Proof. For each B with $x \in -B$ we have $x \in A - B$. Hence, by definition, near(x, A). ∎

Corollary 2.1.9 *If $x = y$, then near$(x, \{y\})$.*

Proposition 2.1.10 *If near(x, A) and $A \subset B$, then near(x, B).*

Proof. This follows directly from the definition of the nearness predicate. ∎

The next result generalises its predecessor and shows that a pre-apartness space satisfies the classical Lodato condition (page 21).

Proposition 2.1.11 *Suppose that* near (x, A), *and that* near(y, B) *for each* $y \in A$. *Then* near(x, B).

Proof. Let $x \in -S$. Then, by the definition of *near*, there exists $y \in A - S$. By hypothesis, near(y, B). Since also $y \in -S$, it follows from the definition of *near* that there exists $z \in B - S$. Thus

$$\forall_{S \subset X} (x \in -S \Rightarrow \exists_{z \in X} (z \in B - S))$$

—that is, near(x, B). ∎

Proposition 2.1.12 *If* near(x, A), *then* near$(x, A \cup B)$ *for all* $B \subset X$.

Proof. Apply Proposition 2.1.10 with B replaced by $A \cup B$. ∎

Proposition 2.1.13 *If* near(x, A) *and* $x \bowtie B$, *then* near$(x, A - B)$.

Proof. Let $x \in -S$. We need to show that $(A - B) - S$ is inhabited. To this end, observe that $x \in -B$, so by **A3**, $x \in -(B \cup S)$. Since near (x, A), there exists $y \in A - (B \cup S)$; but $A - (B \cup S) = (A - B) - S$, so we are through. ∎

One of the axioms of the classical theory of proximity presented in [35] is

$$\text{near}(x, A \cup B) \Leftrightarrow \text{near}(x, A) \lor \text{near}(x, B). \tag{2.9}$$

The implication from right to left here is a consequence of Proposition 2.1.12. To see that the implication from left to right in (2.9) is essentially nonconstructive, consider the metric apartness space $\mathbf{R}$. Given an increasing binary sequence $(a_n)_{n \geqslant 1}$ with $a_1 = 0$, define

$$S \equiv \left\{ \frac{1}{n} : a_n = 0 \right\}, \quad T \equiv \left\{ \frac{1}{n} : a_n = 1 \right\}.$$

Then 0 is near $S \cup T$. But if 0 is near S, then $a_n = 0$ for all n; while if 0 is near T, then there exists $x \in T$ such that $|x| < 1/2$, so we can find n with $a_n = 1$. Thus the left-to-right implication in (2.9) implies **LPO**.

It readily follows from Proposition 2.1.7 that in the context of a metric space, near (x, A) implies $\neg (x \bowtie A)$. This holds in a general pre-apartness space.

Proposition 2.1.14 $\neg (\text{near}(x, A) \land x \bowtie A)$.

Proof. Assume that near(x, A) and $x \bowtie A$. Then $x \in -A$, and so, by the definition of *near*, there exists $y \in A - A$. This contradicts axiom **A2**. ∎

We observe that in the metric apartness space **R**, the statement

$$\forall_{A \subset \mathbf{R}} \left(\neg (1 \bowtie A) \Rightarrow \text{near} (1, A) \right)$$

implies the law of excluded middle in the form

$$\neg\neg P \Rightarrow P.$$

Indeed, let P be any proposition satisfying $\neg\neg P$, and consider the subset

$$A \equiv \{x \in \mathbf{R} : x = 0 \vee (x = 1 \wedge P)\}$$

of **R**. If $1 \bowtie A$, then $\neg P$, which contradicts our initial assumption; hence $\neg (1 \bowtie A)$. But if near $(1, A)$, then there exists $x \in A$ such that $x > 0$; whence $x = 1$, $1 \in A$, and therefore P holds.

Proposition 2.1.15 *If X is a pre-apartness space with the reverse Kolmogorov property, and if near(x, A), and $y \bowtie A$, then $x \neq y$.*

Proof. Assume that near(x, A) and $y \bowtie A$. Proposition 2.1.14 shows that $x \notin -A$; whence $x \neq y$, by (2.4). ∎

Proposition 2.1.16 *If X satisfies **A5** and if $\neg (x \neq y)$, then near$(x, \{y\})$.*

Proof. By **A5**, for each S with $x \in -S$ we have either $x \neq y$ or else $y \in -S$; the former alternative is ruled out, so we must have $y \in -S$. It follows that near $(x, \{y\})$. ∎

An immediate consequence of this proposition is

Corollary 2.1.17 *If X is an apartness space such that*

$$\forall_{x,y \in X} (\text{near}(x, \{y\}) \Rightarrow x = y), \tag{2.10}$$

then the inequality on X is tight.

A pre-apartness space X is said to be $\mathbf{T}_1$ if

$$\forall_{x,y \in X} (x \neq y \Rightarrow x \bowtie \{y\}). \tag{2.11}$$

Since the inequality relation is symmetric, a $\mathbf{T}_1$ pre-apartness space has symmetric point-point pre-apartness:

$$\forall_{x,y \in X} (x \bowtie \{y\} \Leftrightarrow y \bowtie \{x\}).$$

Every quasi-metric apartness space is $\mathbf{T}_1$.

Proposition 2.1.18 If X is a $\mathbf{T}_1$ pre-apartness space and near $(x, \{y\})$ in X, then $\neg\,(x \neq y)$.

Proof. Assume that $x \neq y$; then $x \bowtie \{y\}$, by (2.11). This contradicts Proposition 2.1.14. ∎

Corollary 2.1.19 If X is a $\mathbf{T}_1$ apartness space and near$(x, \{y\})$ in X, then near$(y, \{x\})$.

Proof. Apply Propositions 2.1.18 and 2.1.16, noting that $\neg\,(x \neq y)$ if and only if $\neg\,(y \neq x)$. ∎

Corollary 2.1.20 If X is a $\mathbf{T}_1$ pre-apartness space with tight inequality, then (2.10) holds.

Proof. Apply Proposition 2.1.18. ∎

Proposition 2.1.21 If X is a $\mathbf{T}_1$ apartness space, and a,b are points of X with $a \neq b$, then for each $x \in X$ either $x \neq a$ or $x \neq b$.

Proof. By (2.11), $a \bowtie \{b\}$; whence, by **A5**, either $x \neq a$ or else $x \bowtie \{b\}$; in the latter event, **A2** shows that $x \neq b$. ∎

Proposition 2.1.22 If X is a $\mathbf{T}_1$ apartness space containing two distinct points, then for each $x \in X$ there exists $y \in X$ such that $x \neq y$.

Proof. Let a,b be points of X with $a \neq b$. Then either $x \neq a$ or $x \neq b$, by the previous proposition. ∎

Proposition 2.1.23 If X is a $\mathbf{T}_1$ pre-apartness space, near (x, A), and $x \neq y$, then there exists $z \in A$ such that $y \neq z$.

Proof. In this case, $x \in -\{y\}$, so by the definition of near, there exists $z \in A$ with $z \bowtie \{y\}$. The desired conclusion follows from **A2**. ∎

We end the section by establishing the extensionality of apartness and nearness.

Proposition 2.1.24 If X is an apartness space, $x \bowtie A$, $x = x'$, and $A = A'$, then $x' \bowtie A'$.

Proof. By **A5**, either $x \neq x'$ or else, as must be the case, $x' \bowtie A$. Since $A' = A$, we have $x' \in -A \subset \sim A = \sim A'$, by **A2**; whence $x' \bowtie A'$, by **A4**. ∎

Proposition 2.1.25 If X is an apartness space, near(x, A), $x = x'$, and $A = A'$, then near(x', A').

Proof. Let $x' \in -S$. Then $x \in -S$, by the previous proposition. Since x is near A, there exists $y \in A - S$. But $A' = A$, so $y \in A' - S$. It follows from our definition of near that near(x', A'). ∎

2.2   Apartness and Topology

We assume that every topological space (X, τ) comes equipped with an inequality $\neq$. For a point x and a subset A of such a space we define

$$x \bowtie_\tau A \Leftrightarrow \exists_{U \in \tau} \, (x \in U \subset {\sim}A) \qquad (2.12)$$

and, of course,

$$\mathrm{near}(x, A) \Leftrightarrow \forall_{B \subset X} \, (x \in -B \Rightarrow \exists_{y \in X} \, (y \in A - B)) \, .$$

Thus $x \bowtie_\tau A$ if and only if x belongs to $({\sim}A)^\circ$, the interior of ${\sim}A$ relative to the topology τ. It is easy to show that $\bowtie_\tau$ satisfies **A1–A4** for a pre-apartness— the **topological pre-apartness** corresponding to τ. We call X, taken with this pre-apartness, a **topological pre-apartness space**, or, if $\bowtie_\tau$ also satisfies **A5**, a **topological apartness space**. It is easily shown that if a topological space (X, τ) has the **topological A5 property**

$$\forall_{x \in X} \forall_{U \in \tau} \, (x \in U \Rightarrow \forall_{y \in X} \, (x \neq y \vee y \in U)) \, , \qquad (2.13)$$

then the corresponding pre-apartness satisfies **A5**. Note that on a quasi-metric apartness space the pre-apartness corresponding to the quasi-metric topology coincides with the quasi-metric apartness.

The alert reader may have asked: is a point-set pre-apartness necessarily symmetric—that is, do we have

$$\forall_{x,y \in X} \, (x \bowtie \{y\} \Rightarrow y \bowtie \{x\})?$$

We give a topological example to show that it is not. Consider the set $X \equiv \{0, 1\}$ with the discrete equality and inequality, the topology

$$\tau \equiv \{\varnothing, \{1\}, \{0, 1\}\} \, ,$$

and the corresponding topological pre-apartness $\bowtie_\tau$. Since $1 \in \{1\} \subset {\sim}\{0\}$, we have $1 \bowtie_\tau \{0\}$; but since $0 \notin \{1\}$ and $\{0, 1\} \not\subset {\sim}\{1\}$, we cannot have $0 \bowtie_\tau \{1\}$.

Occasionally it is helpful to consider the following condition, linking the inequality and the topology, on a topological space (X, τ):

$$\forall_{x,y \in X} \forall_{U \in \tau} \, ((x \in U \wedge y \notin U) \Rightarrow x \neq y) \, . \qquad (2.14)$$

This condition always holds classically for the denial inequality, and is a constructive consequence of the topological **A5** property. Thus it holds if the space is discrete, and in any quasi-metric space.

Proposition 2.2.1 *Let (X, τ) be a topological space in which condition* (2.14) *holds. Then $\bowtie_\tau$ has the reverse Kolmogorov property.*

Proof. Let $x \in -S$ and $y \notin -S$ in X. Then there exists $U \in \tau$ such that $x \in U \subset \sim S$. It follows from this and the definition of $\bowtie_\tau$ that $U \subset -S$, so $y \notin U$; whence $x \neq y$, by (2.14). ∎

In light of Proposition 2.2.1, we refer to condition (2.14) as the **topological reverse Kolmogorov property** (with *topological* omitted when it is clear that we are referring to (2.14) rather than its counterpart (2.4) for an apartness space).

The following proposition will re-appear in Chapter 3.

Proposition 2.2.2 *Let (X, τ) be a topological space with the topological reverse Kolmogorov property. Then $(\neg S)^\circ = (\sim S)^\circ$ for each subset S of X.*

Proof. It is trivial that $(\sim S)^\circ \subset (\neg S)^\circ$. For the reverse inclusion, let $x \in (\neg S)^\circ$ and $y \in S$. Clearly, $y \notin (\neg S)^\circ$. Since $(\neg S)^\circ \in \tau$, it follows from (2.14) that $x \neq y$. Thus $(\neg S)^\circ \subset \sim S$. Since also $(\neg S)^\circ$ is open, it follows that $(\neg S)^\circ \subset (\sim S)^\circ$. ∎

Proposition 2.2.3 *Let (X, τ) be a topological space with the topological reverse Kolmogorov property, let Y be a subspace of X, and let τ_Y be the subspace topology induced on Y by τ. Then the topological subspace (Y, τ_Y) also has the topological reverse Kolmogorov property.*

Proof. Let $x, y \in Y$ and $V \in \tau_Y$ be such that $x \in V$ and $y \notin V$. By the definition of τ_Y, there exists $U \in \tau$ such that $V = Y \cap U$. If $y \in U$, then $y \in Y \cap U = V$, a contradiction; hence $y \notin U$. Since $x \in U$, it follows from condition (2.14) in X that $x \neq y$. ∎

A topological pre-apartness is $\mathbf{T}_1$ (as defined at (2.11) on page 26) if and only if

$$\forall_{x,y \in X} (x \neq y \Rightarrow \exists_{U \in \tau} (x \in U \wedge y \in \sim U))$$

—in other words, its topology τ satisfies the classical **first axiom of separation**. A quasi-metric space is a $\mathbf{T}_1$ topological apartness space.

Another example of a $\mathbf{T}_1$ apartness space is that associated with a **locally convex space**—that is, a vector space X over $\mathbf{R}$ or $\mathbf{C}$, together with a family $(p_i)_{i \in I}$ of seminorms that satisfies the following condition:

$$\forall_{x,y \in X} (x \neq y \Leftrightarrow \exists_{i \in I} (p_i(x - y) > 0)).$$

Recall from page 15 that the inequality on a vector space is tight, by definition; whence we must have

$$\forall_{x,y \in X} (x = y \Leftrightarrow \forall_{i \in I} (p_i(x - y) = 0)).$$

The family $(p_i)_{i \in I}$ defines the corresponding **locally convex topology**, in which the basic open neighbourhoods of x are of the form

$$\{x' \in X : \forall_{i \in F} (p_i(x - x') < \varepsilon)\}$$

for some $\varepsilon > 0$ and some inhabited, finitely enumerable subset F of I. It is left as an exercise for the reader to show that this topology has the reverse Kolmogorov property (2.14). To establish **A5** for the corresponding topological pre-apartness $\bowtie$, we argue as follows. Let $x \bowtie A$; then there exist $\varepsilon > 0$ and elements $i_1,\ldots,i_n$ of I such that

$$U \equiv \left\{ y \in X : \sum_{k=1}^{n} p_{i_k}(x - y) < \varepsilon \right\} \subset \sim A.$$

Given $y \in X$, we have either $\sum_{k=1}^{n} p_{i_k}(x - y) > 0$ or $\sum_{k=1}^{n} p_{i_k}(x - y) < \varepsilon$. In the first case, $p_{i_k}(x - y) > 0$ for some k and so $x \neq y$. In the second case, $y \in U \subset \sim A$ and so, by definition of the pre-apartness, $y \bowtie A$. This completes the proof of **A5**. It is easy to see that the topological pre-apartness relation $\bowtie$ is $\mathbf{T}_1$.

If (X,τ) is a topological space, and Y an inhabited subset of X, then the subspace topology τ_Y induced on Y by τ has an associated pre-apartness $\bowtie_{\tau_Y}$. A natural question arises: under what conditions does this pre-apartness coincide with the restriction of the relation $\bowtie_\tau$ to points and subsets of Y?

Proposition 2.2.4 *Let X be a topological space satisfying (2.13), and let $\bowtie_\tau$ be the corresponding topological apartness. Let Y be an inhabited subset of X, let τ_Y be the subspace topology induced on Y by τ, and let $\bowtie_{\tau_Y}$ be the corresponding topological apartness on Y. Then*

$$\forall_{y \in Y} \forall_{S \subset Y} \, (y \bowtie_{\tau_Y} S \Leftrightarrow y \bowtie_\tau S).$$

Proof. Let $y \in Y \subset X$ and $S \subset Y$. If $y \bowtie_\tau S$, then there exists $U \in \tau$ such that $y \in U \subset X \sim S$; whence

$$y \in Y \cap U \subset Y \cap (X \sim S) = Y \sim S,$$

where $Y \cap U \in \tau_Y$; so $y \bowtie_{\tau_Y} S$.

Conversely, if $y \bowtie_{\tau_Y} S$, then there exists $V \in \tau_Y$ such that $y \in V \subset Y \sim S$. By definition of the subspace topology τ_Y, there exists $U \in \tau$ such that $V = Y \cap U$. Hence

$$y \in Y \cap U \subset Y \sim S \subset X \sim S.$$

Consider any $x \in U$ and any $s \in S$. By (2.13), either $x \neq s$ or $s \in U$; in the latter case, $s \in Y \cap U$, so $s \in Y \sim S$, which is absurd. It follows that $y \in U \subset X \sim S$ and therefore $y \bowtie_\tau S$. ∎

Corollary 2.2.5 *Let Y be an inhabited subset of a topological space (X,τ) that satisfies (2.13). Then the restriction of the relation $\bowtie_\tau$ to points and subsets of Y is an apartness on Y.*

Proof. According to Proposition 2.2.4, the restriction in question is precisely the apartness induced on Y by the subspace topology τ_Y. ∎

Corollary 2.2.6 *Let Y be an inhabited subset of a quasi-metric space X. Then the apartness induced on Y by the quasi-metric on X coincides with the apartness induced on Y by the original quasi-metric apartness on X.*

We now introduce an important topology on a given pre-apartness space $(X,\bowtie)$. A subset S of X is said to be **nearly open** if it can be written as a union of apartness complements: that is, if there exists a family $(A_i)_{i\in I}$ of subsets of X such that $S = \bigcup_{i\in I} -A_i$. The empty subset of X is nearly open ($\varnothing = -X$), X is nearly open ($X = -\varnothing$), and a union of nearly open sets is nearly open. Since, by a simple induction argument using **A3**, the intersection of a finite number of apartness complements is an apartness complement, it can easily be shown that a finite intersection of nearly open sets is nearly open. Thus the nearly open sets form a topology—the **apartness topology**, denoted by $\tau_\bowtie$—on X for which the apartness complements form a basis. It is straightforward to prove that if $(X,\bowtie)$ is an apartness space, then the corresponding apartness topology satisfies (2.13).

Proposition 2.2.7 *In a topological pre-apartness space every nearly open set is open.*

Proof. Let (X,τ) be a topological pre-apartness space. It suffices to show that every apartness complement $-A$ in X is open. Let $x \in -A$ and choose $U \in \tau$ such that $x \in U \subset \sim A$. Then, by the definition of the apartness in X, $U \subset -A$. Hence $-A$ is open. ∎

Corollary 2.2.8 *Let $(X,\bowtie)$ be a pre-apartness space, and let τ be the corresponding apartness topology. Then the topological pre-apartness structure corresponding to τ coincides with the original pre-apartness structure on X.*

Proof. Let $\bowtie_\tau$ denote the topological apartness corresponding to τ. If $x \bowtie S$, then $x \in -S \subset \sim S$, where $-S$ is the apartness complement of S relative to $\bowtie$; since $-S$ is open by Proposition 2.2.7, it follows from the definition of the topological pre-apartness that $x \bowtie_\tau S$.

Conversely, if $x \bowtie_\tau S$, then there exists a nearly open set U such that $x \in U \subset \sim S$, so (by definition of *nearly open*) there exists V such that $x \in -V \subset \sim S$. It follows from **A4** that $x \bowtie S$. ∎

Corollary 2.2.8 provides the motivation for the definition (in Section 2.5 below) of the product of two pre-apartness spaces.

Proposition 2.2.9 *If X is a pre-apartness space with the reverse Kolmogorov property, then the corresponding apartness topology has the topological reverse Kolmogorov property.*

Proof. Consider $x, y \in X$ and $U \in \tau_{\bowtie}$ such that $x \in U$ and $y \notin U$. There exists $S \subset X$ such that $x \in -S \subset U$. Clearly, $y \notin -S$, so, by the reverse Kolmogorov property in $(X, \bowtie)$, we have $x \neq y$. ∎

We say that a topological pre-apartness space is **topologically consistent** if every open subset of X is nearly open.

Proposition 2.2.10 *Every metric apartness space is topologically consistent.*

Proof. Let (X, ρ) be a metric space, and let $\bowtie$ be the corresponding metric apartness, which, as noted earlier, coincides with the topological apartness corresponding to the metric topology on X. In view of Proposition 2.2.7, it suffices to prove that every (metrically) open subset S of X is nearly open. Given $x \in S$, choose $r > 0$ such that $B(x, r) \subset S$. Then $x \in -\sim B(x, r/2) \subset B(x, r) \subset S$. It follows that x is an interior point of S in the apartness topology. Hence S is nearly open. ∎

Proposition 2.2.11 *The following conditions are equivalent on a topologically consistent topological space (X, τ).*

(i) (X, τ) *has the topological reverse Kolmogorov property* (2.14).

(ii) $(X, \bowtie_\tau)$ *has the reverse Kolmogorov property* (2.4).

Proof. Since τ coincides with the apartness topology arising from $\bowtie_\tau$, the desired conclusion follows from Propositions 2.2.1 and 2.2.9. ∎

Classically, for every open set A of a topological pre-apartness space we have $A = \sim \sim A$ and therefore, by the definition of the topological pre-apartness, $A = -\sim A$, from which it follows that the space is topologically consistent. Constructively, although $A \subset -\sim A$ holds for an open set A, we cannot hope to prove the reverse inclusion even in the presence of (2.14). To see this, consider the metric subspace

$$X \equiv \{0, 1 - a, 2\}$$

of $\mathbf{R}$, where $a < 1$ and $\neg(a \leqslant 0)$. Let B be the open ball with centre 0 and radius 1 in X. Then $2 \in \sim B$. On the other hand, if $x \in \sim B$ and $x \neq 2$, then x must equal $1 - a$, so $\neg(1 - a < 1)$; whence $1 - a \geqslant 1$ and therefore $a \leqslant 0$, a contradiction. It follows that $\sim B = \{2\}$ and hence that $-\sim B = -\{2\}$. If $-\sim B \subset B$, then $1 - a$, which certainly belongs to $-\{2\}$, is in B; whence $1 - a < 1$ and therefore $a > 0$. Thus, although (as observed in the proof of Proposition 2.2.10) any open ball B in a metric apartness space satisfies $B \subset -\sim B$, the proposition

Every open ball B in a metric space satisfies $B = -\sim B$

entails

$$\forall_{x \in \mathbf{R}} \left(\neg \left(x \leqslant 0 \right) \Rightarrow x > 0 \right),$$

a statement equivalent to Markov's principle. In fact, as is shown by an example which we now present, if every $\mathbf{T}_1$ topological apartness space with discrete inequality is topologically consistent, then the law of excluded middle holds.

Let A be the set of odd positive integers, B the set of even positive integers, and X the space $\mathbf{N}$, taken with the denial inequality. Given a statement P, for each positive integer n let

$$U_n \equiv \{0\} \cup \{k \in A : k > n\} \cup \{k \in B : k > n \wedge P\}.$$

The sets U_n, together with the singletons $\{n\}$ with $n \in \mathbf{N}^+$, form a countable basis for a **second-countable topology** τ (that is, a topology with a countable base of open sets) on X. Note that any neighbourhood of 0 in this topology must contain one of the sets U_n. Since X is discrete, it automatically satisfies **A5**; so τ induces a topological apartness structure on X. Suppose that X is topologically consistent. Since U_1 is open, it is nearly open and therefore there exists $V \subset X$ such that $0 \in -V \subset U_1$. By Proposition 2.2.7, $-V$ is τ-open, so there exists N such that $0 \in U_N \subset -V$. Assume that $\neg\neg P$ holds. If $V \cap \{k \in B : k > N\}$ is inhabited, then since $U_N \subset -V$, we must have $\neg P$, a contradiction. Hence

$$\{k \in B : k > N\} \subset \neg V.$$

In view of the discreteness of X, we have $\neg V = {\sim} V$. On the other hand, given $m \in {\sim} V$, we have either $m = 0 \in -V$ (by our choice of V), or else $m \geqslant 1$, $m \in \{m\} \subset {\sim} V$, and therefore $m \bowtie V$, by definition of the topological pre-apartness. It follows from this and axiom **A2** that ${\sim} V = -V$; whence

$$\{k \in B : k > N\} \subset -V \subset U_1,$$

which is possible if and only if P holds. We conclude that if every topological apartness space is topologically consistent, then the law of excluded middle holds.

Note that in the foregoing example the space X is $\mathbf{T}_1$. Indeed, if $n \in X$ and $n \neq 0$, then $0 \in U_{n+1}$ and $n \in {\sim} U_{n+1}$; whereas if $m, n \in X$ and $n > m \geqslant 1$, then $n \in \{n\}, m \in \{m\}$, and $\{n\} \subset {\sim} \{m\}$.

As we shall see in a moment, the following property ensures topological consistency. We say that a topological space (X, τ), or just the topology τ itself, is **topologically locally decomposable** if

$$\forall_{x \in X} \forall_{U \in \tau} \left(x \in U \Rightarrow \exists_{V \in \tau} \left(x \in V \wedge X = U \cup {\sim} V \right) \right).$$

This condition holds classically for any topological space with the denial inequality: just take $V = U$. Every metric space (X, ρ) is topologically locally decomposable: for if $x \in U$ and U is open in X, then, choosing $r > 0$ such that $B(x, r) \subset U$, we can take $V = B(x, r/2)$.

Proposition 2.2.12 *A topologically locally decomposable topological preapartness space* (X,τ) *is topologically consistent.*

Proof. Given $U \in \tau$ and $x \in U$, find $V \in \tau$ such that $x \in V$ and $X = U \cup {\sim}V$. Then $x \in V \subset {\sim}{\sim}V$, so $x \in -{\sim}V$. Since

$$-{\sim}V \subset {\sim}{\sim}V \subset U,$$

it follows that U is a union of apartness complements and so is nearly open. ∎

We can see from Proposition 2.2.12 and the foregoing Brouwerian example that if every topological apartness space is topologically locally decomposable, then the law of excluded middle holds.

How do we produce a property like topological local decomposability but applicable to arbitrary pre-apartness spaces? We say that a pre-apartness space $(X,\bowtie)$, or just the pre-apartness $\bowtie$ itself, is **locally decomposable** if

$$\forall_{x \in X} \forall_{S \subset X} \left(x \in -S \Rightarrow \exists_{T \subset X} \left(x \in -T \wedge X = -S \cup T \right) \right).$$

Local decomposability always holds classically: for if $x \in -S$, then, taking $T = {\sim} - S$, we have $X = -S \cup T$; also, by Proposition 2.1.5,

$$-S = -{\sim} - S = -T,$$

so $x \in -T$. Every metric space (X,ρ) is locally decomposable: for if $x \in -S$, then, choosing $r > 0$ such that $B(x,r) \subset -S$, we can take $T \equiv {\sim}B(x,r/2)$ to obtain $x \in -T$ and $X = -S \cup T$.

Proposition 2.2.13 *A locally decomposable pre-apartness space satisfies* **A5**.

Proof. Let $(X,\bowtie)$ be locally decomposable, let $x \in -S$, and choose T such that $x \in -T$ and $X = -S \cup T$. For each $y \in X$, either $y \in -S$ or $y \in T$; in the latter case, $x \bowtie \{y\}$ by Proposition 2.1.1, and so, by **A2**, $x \neq y$. ∎

It will be clear as we develop the subject that local decomposability, which gives us not only **A5** but also stronger alternative conditions to play with, is a very powerful condition.

Proposition 2.2.14 *Let* $(X,\bowtie)$ *be a locally decomposable apartness space, and* τ *the corresponding apartness topology. Then* (X,τ) *is topologically locally decomposable.*

Proof. Let $U \in \tau$ and $x \in U$. Without loss of generality, we may assume that $U = -S$ for some $S \subset X$. Choosing $T \subset X$ such that $x \in -T$ and $X = -S \cup T$, set $V \equiv -T$; then $x \in V$. Moreover, for each $y \in X$ either $y \in -S = U$ or else

$$y \in T \subset {\sim}{\sim}T \subset {\sim} - T = {\sim}V.$$

Since x and U are arbitrary, we conclude that (X, τ) is topologically locally decomposable. ∎

Proposition 2.2.15 *The induced pre-apartness on a topologically locally decomposable topological space is locally decomposable.*

Proof. Let (X, τ) be a topologically locally decomposable topological space. Let $\bowtie_\tau$ denote the corresponding topological apartness, and $-_\tau$ the corresponding operation of apartness complementation. Given $x \in -_\tau S$, and noting that $-S$ is in τ (by Proposition 2.2.7), construct $V \in \tau$ such that $x \in V$ and $X = -S \cup \sim V$. Let $T \equiv \sim V$. Then $X = -S \cup T$. Moreover, $x \in V \subset \sim T$, so $x \bowtie_\tau T$. Since x and S are arbitrary, it follows that $(X, \bowtie_\tau)$ is locally decomposable. ∎

Corollary 2.2.16 *Let (X, τ) be a topological pre-apartness space. Then the following conditions are equivalent.*

(i) *X is topologically locally decomposable.*

(ii) *X is topologically consistent, and $(X, \bowtie_\tau)$ is locally decomposable.*

Proof. Assuming (i), we see from Proposition 2.2.12 that X is topologically consistent, and from Proposition 2.2.15 that it is locally decomposable. On the other hand, if (ii) holds, then Proposition 2.2.14 shows that $(X, \tau_\bowtie)$ is topologically locally decomposable; so if X is also topologically consistent, then (i) holds. ∎

In view of Corollary 2.2.16, for a topologically consistent topological pre-apartness space there should be no confusion if we use the phrases *topologically locally decomposable* and *locally decomposable* interchangeably.

Nearness now comes into play again, in our discussion of closed sets in the apartness topology.

Proposition 2.2.17 *Let A be a subset of the pre-apartness space X. Then*

$$\overline{A} = \{x \in X : \text{near}(x, A)\},$$

where the bar denotes closure in the apartness topology.

Proof. Let near(x, A), and let $U \equiv \bigcup_{i \in I} -A_i$ be any nearly open set containing x. Choosing $i \in I$ such that $x \in -A_i$, we see from Proposition 2.1.13 that near$(x, A - A_i)$. So, by Proposition 2.1.6, there exists $y \in A - A_i \subset A \cap U$. Conversely, if A intersects each nearly open set containing x, then since $-B$ is nearly open for each $B \subset X$, we see immediately from the definition of *near* that near(x, A). ∎

Corollary 2.2.18 *A subset A of a pre-apartness space is closed in the apartness topology if and only if it is **nearly closed**, in the sense that*

$$A = \{x \in X : \text{near}(x, A)\}.$$

Proposition 2.2.19 *For each nearly open subset of a pre-apartness space X whose apartness topology has the reverse Kolmogorov property, the logical complement equals the complement and is nearly closed.*

Proof. Let $A \equiv \bigcup_{i \in I} -U_i$ be nearly open, and $x \in \overline{\neg A}$. Then near$(x, \neg A)$, by Proposition 2.2.17. For each $i \in I$ it follows that if $x \in -U_i$, then there exists $z \in (\neg A) - U_i \subset (\neg A) \cap A$, which is absurd. We conclude that $x \notin -U_i$; whence $x \in \sim - U_i$, by the reverse Kolmogorov property. Thus

$$x \in \bigcap_{i \in I} \sim - U_i = \sim A.$$

Hence

$$\neg A \subset \overline{\neg A} \subset \sim A \subset \neg A,$$

from which the desired conclusions now follow. ∎

Having established the fundamental results connecting pre-apartness and topologies, we turn in the next section to an examination of types of continuity of mappings between pre-apartness spaces.

2.3 Apartness and Continuity

Intuitionistic logic enables us to distinguish between various classically equivalent types of continuity that we now introduce.

Let $f : X \to Y$ be a mapping between pre-apartness spaces. We say that f is

▷ **nearly continuous** if

$$\forall_{x \in X} \forall_{A \subset X} \; (\text{near}(x, A) \Rightarrow \text{near}(f(x), f(A)));$$

▷ **continuous** if

$$\forall_{x \in X} \forall_{A \subset X} \; (f(x) \bowtie f(A) \Rightarrow x \bowtie A)$$

—that is,

$$f^{-1}(-f(A)) \subset -A$$

for each $A \subset X$;

▷ **topologically continuous** if $f^{-1}(S)$ is nearly open in X for each nearly open $S \subset Y$.

It is almost trivial that the composition of continuous functions is continuous, and that the restriction of a continuous function to a pre-apartness subspace of its domain is continuous. Analogous remarks hold for nearly continuous functions and for topologically continuous ones.

For a mapping between quasi-metric spaces, continuity in our sense turns out to be equivalent to the standard one from elementary analysis courses.

Proposition 2.3.1 *The following are equivalent conditions on a mapping $f : X \to Y$ between quasi-metric spaces.*

(i) *f is continuous.*

(ii) *For each $x \in X$ and each $\varepsilon > 0$, there exists $\delta > 0$ such that if $x' \in X$ and $\rho(x,x') < \delta$, then $\rho(f(x),f(x')) < \varepsilon$.*

Proof. Suppose that f is continuous. Given $x \in X$ and $\varepsilon > 0$, let

$$A \equiv \left\{ x' \in X : \rho(f(x), f(x')) > \frac{\varepsilon}{2} \right\}.$$

Then $f(x) \bowtie f(A)$ in Y, so $x \bowtie A$ in X. Pick $\delta > 0$ such that $\rho(x,x') \geqslant \delta$ for all $x' \in A$. If $x' \in X$ and $\rho(x,x') < \delta$, then $x' \notin A$ and therefore $\rho(f(x),f(x')) < \varepsilon$. Thus (i) implies (ii).

Now suppose that (ii) holds, and let $f(x) \bowtie f(A)$ in Y. Then there exists $\varepsilon > 0$ such that $\rho(f(x),f(x')) \geqslant \varepsilon$ for all $x' \in A$. Pick $\delta > 0$ as in (ii). If $x' \in A$ and $\rho(x,x') < \delta$, then $\rho(f(x),f(x')) < \varepsilon$, a contradiction. Hence $\rho(x,x') \geqslant \delta$ for each $x' \in A$; in other words, $x \bowtie A$. Hence (ii) implies (i). ∎

A continuous mapping f of a pre-apartness space into a $\mathbf{T}_1$ pre-apartness space is strongly extensional: for if $f(x) \neq f(y)$, we have $f(x) \bowtie \{f(y)\}$; whence $x \bowtie \{y\}$ and therefore, by **A2**, $x \neq y$. Less trivial to establish is the strong extensionality of nearly continuous mappings.

Proposition 2.3.2 *A nearly continuous mapping of an apartness space into a $\mathbf{T}_1$ pre-apartness space is strongly extensional.*

Proof. Let $f : X \to Y$ be nearly continuous, where X is an apartness space and Y is a $\mathbf{T}_1$ pre-apartness space. Consider $x,x' \in X$ such that $f(x) \neq f(x')$. Define
$$A \equiv \{z \in X : z = x \vee (z = x' \wedge x \neq x')\}.$$
Note that $x \in A$. Consider any $U \subset X$ such that $x' \in -U$; by **A5**, either $x \neq x'$ and therefore $x' \in A - U$, or else $x \in -U$ and so $x \in A - U$. It follows that near(x', A). Using the near continuity of f, we obtain near$(f(x'),f(A))$. Since also $f(x) \neq f(x')$, Proposition 2.1.23 shows there exists $z \in A$ such that $f(z) \neq f(x)$. Then $\neg(z = x)$, so we must have $z = x'$ and $x \neq x'$. ∎

Proposition 2.3.3 *The following conditions are equivalent on a mapping $f : X \to Y$ between pre-apartness spaces.*

(i) *f is nearly continuous.*

(ii) *For each nearly closed subset S of Y, $f^{-1}(S)$ is nearly closed.*

(iii) *For each subset A of X, $f(\overline{A}) \subset \overline{f(A)}$.*

Proof. Suppose that f is nearly continuous on X, and let S be a nearly closed subset of Y. If $x \in \overline{f^{-1}(S)}$, then near $(x, f^{-1}(S))$, by Proposition 2.2.17, and therefore near$(f(x), S)$. Since S is nearly closed, $f(x) \in S$; whence $x \in f^{-1}(S)$. Thus (i) implies (ii).

Now suppose that (ii) holds. Let $x \in X$ and $A \subset X$ be such that near(x, A). Note that $A \subset f^{-1}\left(\overline{f(A)}\right)$, so near$\left(x, f^{-1}(\overline{f(A)})\right)$, by Proposition 2.1.10. Since, by Corollary 2.2.18, $\overline{f(A)}$ is nearly closed, so is $f^{-1}(\overline{f(A)})$. Hence $x \in f^{-1}\left(\overline{f(A)}\right)$, so $f(x) \in \overline{f(A)}$ and therefore near$(f(x), f(A))$, again by Proposition 2.2.17. Thus (ii) implies (i).

The equivalence of (i) and (iii) is a consequence of Proposition 2.2.17. ∎

Proposition 2.3.4 *A topologically continuous mapping $f : X \to Y$ between pre-apartness spaces is nearly continuous.*

Proof. Consider $x \in X$ and $A \subset X$ such that near(x, A). Let $B \subset Y$ and $f(x) \in -B$; then $x \in f^{-1}(-B)$. By the topological continuity of f, there exists a family $(A_i)_{i \in I}$ of subsets of X such that $f^{-1}(-B) = \bigcup_{i \in I} -A_i$. Choose i_0 with $x \in -A_{i_0}$. Since near(x, A), there exists

$$y \in A - A_{i_0} \subset A \cap \left(\bigcup_{i \in I} -A_i\right);$$

whence

$$f(y) \in f(A) \cap f\left(\bigcup_{i \in I} -A_i\right) \subset f(A) - B.$$

Since B is arbitrary, we conclude that near$(f(x), f(A))$. ∎

Corollary 2.3.5 *Every topologically continuous mapping of an apartness space into a T_1 pre-apartness space is strongly extensional.*

Proof. Apply Propositions 2.3.4 and 2.3.2. ∎

Proposition 2.3.6 *Let X be a pre-apartness space whose apartness topology has the reverse Kolmogorov property, and let f be a topologically continuous mapping of X into a pre-apartness space Y. Then f is continuous.*

Proof. Given $A \subset X$ and writing

$$B \equiv f^{-1}(-f(A)),$$

we see that $A \cap B = \varnothing$ and, by topological continuity, that $B = \bigcup_{i \in I} -V_i$ for some family $(V_i)_{i \in I}$ of subsets of X. It follows from Proposition 2.2.19 that $\neg B = {\sim}B$. For each i we therefore have

$$A \subset \neg B = {\sim}B \subset {\sim} - V_i$$

and therefore

$$-V_i \subset \sim\sim -V_i \subset \sim A.$$

Applying **A4**, we obtain $-V_i \subset -A$. Hence $B \subset -A$, and therefore f is continuous. ∎

In order to obtain a partial converse to Proposition 2.3.6, we introduce the following **weak nested neighbourhoods property** for a pre-apartness space X.

WNN: $\quad x \in -A \Rightarrow \exists_{B \subset X} (x \in -B \land (\neg B \subset -A))$.

This property is a simple consequence of local decomposability. It captures the idea that inside every basic neighbourhood of a point, relative to the apartness topology, there should be a strictly smaller neighbourhood of that point. In Chapter 3, when we have introduced a notion of apartness between sets, we shall deal with a stronger nested neighbourhoods property than this weak one. For our present purpose, though, the latter is certainly adequate.

Proposition 2.3.7 Let X be a pre-apartness space, and Y a pre-apartness space with the weak nested neighbourhoods property. Then every continuous function $f : X \to Y$ is topologically continuous.

Proof. Let $S \equiv \bigcup_{i \in I} -A_i$ be a nearly open subset of Y, and consider any $x \in f^{-1}(S)$. Choose $i \in I$ such that $f(x) \in -A_i$. By the weak nested neighbourhoods property, there exists $B \subset Y$ such that

$$f(x) \in -B \subset \neg B \subset -A_i.$$

It follows from this and the continuity of f that

$$x \in -f^{-1}(B) \subset f^{-1}(\neg B) \subset f^{-1}(-A_i) \subset f^{-1}(S).$$

Hence $f^{-1}(S)$ is a union of apartness complements in X and is therefore nearly open. ∎

Propositions 2.3.6 and 2.3.7 now yield

Corollary 2.3.8 Let X be f be a mapping of an apartness space X into a pre-apartness space Y with the weak nested neighbourhoods property. Then f is continuous if and only if it is topologically continuous.

We end the section with a general type of topological space that has the weak nested neighbourhoods property. A pre-apartness space X is said to be **completely regular** if for each $x \in X$ and each $A \subset X$ with $x \bowtie A$, there exists a continuous function $\phi : X \to [0,1]$ such that $\phi(x) = 0$ and $\phi(A) \subset \{1\}$. In that case, X is actually an apartness space. Every metric apartness space is completely regular.

Proposition 2.3.9 *A completely regular apartness space has the weak nested neighbourhoods property.*

Proof. Let X be completely regular, and let $x \in -A$ in X. There exists a continuous function $\phi : X \to [0,1]$ such that $\phi(x) = 0$ and $\phi(A) \subset \{1\}$. Let $B \equiv \phi^{-1}\left(\frac{1}{2}, 1\right]$. Since $\rho(\phi(x), \phi(y)) > 1/2$ for each $y \in B$, the continuity of ϕ ensures that $x \in -B$. Moreover, for each $z \in \neg B$ we have $\phi(z) \leqslant 1/2$, so $\phi(z) \bowtie \phi(A)$ and therefore, again by the continuity of ϕ, $z \in -A$. Hence $\neg B \subset -A$. ∎

We see from this and Corollary 2.3.8 that for mappings from an apartness space into a completely regular space, continuity and topological continuity are equivalent.

The following diagram summarises the main connections between types of continuity for functions between pre-apartness spaces.

2.4   Limits

How do we fit convergence and limits into our framework? We first need to introduce nets as a generalisation of sequences.

Let $\mathfrak{D}$ be an inhabited set. By a **preorder** on $\mathfrak{D}$ we mean a binary relation on $\mathfrak{D}$ that is both reflexive,

$$\forall_{x \in \mathfrak{D}} \left(x \succcurlyeq x \right),$$

and transitive,

$$\forall_{x,y,z \in \mathfrak{D}} \left((x \succcurlyeq y \wedge y \succcurlyeq z) \Rightarrow x \succcurlyeq z \right).$$

We call the pair $(\mathfrak{D}, \succcurlyeq)$—or, when it is clear which partial order we are dealing with, $\mathfrak{D}$ itself—a **directed set** if for all $m, n \in \mathfrak{D}$, there exists $p \in \mathfrak{D}$ such that $p \succcurlyeq m$ and $p \succcurlyeq n$. A **net** in a set X is a mapping $n \rightsquigarrow x_n$ of such a set $\mathfrak{D}$—the **index set**—into X, and is normally denoted by $(x_n)_{n \in \mathfrak{D}}$. If f is a mapping of X into a set Y, then $(f(x_n))_{n \in \mathfrak{D}}$ is a net in Y. A sequence in X is just a special case of a net in which the index set is the set $\mathbf{N}^+$.

To each x in a pre-apartness space X there correspond two special nets defined as follows. Let

$$\mathfrak{D}_x \equiv \{(\xi, U) : x \in -U \wedge \xi \in -U\},$$

with equality[2] defined by

$$(\xi, U) = (\xi', U') \Leftrightarrow (\xi = \xi' \wedge -U = -U').$$

For each $n \equiv (\xi, U)$ in $\mathfrak{D}_x$ define $x_n \equiv \xi$. It is easy to see that $\mathfrak{D}_x$ is a directed set under the **reverse inclusion preorder** defined by

$$(\xi, U) \succcurlyeq (\xi', U') \Leftrightarrow -U \subset -U',$$

so that $\mathcal{N}_x \equiv (x_n)_{n \in \mathfrak{D}_x}$ is a net—the **basic neighbourhood net** of x. Similarly,

$$\mathfrak{D}'_x \equiv \{(\xi, U) : x \in -U \wedge \xi \in -U \wedge \xi \neq x\}$$

is a directed set, and $\mathcal{N}'_x \equiv (x_n)_{n \in \mathfrak{D}'_x}$ is a net—the **basic punctured neighbourhood net** of x.

For convenience we introduce here a simple but valuable lemma.

Lemma 2.4.1 Let X be a pre-apartness space, x a point of X, U a subset of X with $x \in -U$, and $\nu \equiv (\xi, U)$. Then

$$-U = \{x_n : n \in \mathfrak{D}_x, n \succcurlyeq \nu\}. \tag{2.15}$$

If near $(x, X \sim \{x\})$, then

$$-U \cap \sim \{x\} = \{x_n : n \in \mathfrak{D}'_x, n \succcurlyeq \nu\}. \tag{2.16}$$

Proof. If $n \equiv (x_n, V) \succcurlyeq \nu$, then $x_n \in -V \subset -U$. Hence

$$\{x_n : n \succcurlyeq \nu\} \subset -U.$$

On the other hand, for each $y \in -U$ we have $(y, U) \succcurlyeq \nu$, so $y \in \{x_n : n \succcurlyeq \nu\}$. Both (2.15) and (2.16) readily follow from this. ∎

In view of the standard elementary notion of convergence of sequences in a metric space and of the definition of the apartness topology, it makes sense to say that a net $(x_n)_{n \in \mathfrak{D}}$ in a pre-apartness space X **converges** to a **limit** x in X if

$$\forall_{U \subset X} (x \in -U \Rightarrow \exists_{N \in \mathfrak{D}} \forall_{n \succcurlyeq N} (x_n \in -U)).$$

We also say that the net is **apartness convergent** to x. It follows from Lemma 2.4.1 that the basic neighbourhood net of a point in a pre-apartness space converges to that point.

---

[2]We do not need to bother with the natural inequality on either $\mathfrak{D}_x$ or $\mathfrak{D}'_x$, so we leave the definition of this to the reader.

Proposition 2.4.2 *Let X be a pre-apartness space with the weak nested neighbourhoods property. Then the net $s \equiv (x_n)_{n \in \mathfrak{D}}$ converges to x in X if and only if*

$$\forall_{B \subset \mathfrak{D}} \left(x \bowtie s(B) \Rightarrow \exists_{N \in \mathfrak{D}} \left(B \subset \neg \{n : n \succcurlyeq N\} \right) \right). \qquad (2.17)$$

Proof. Suppose that s converges to x in X. Let $B \subset \mathfrak{D}$ and $x \bowtie s(B)$. By the weak nested neighbourhoods property, there exists $U \subset X$ such that $x \in -U$ and $\neg U \subset -s(B) \subset \neg s(B)$; whence

$$s(B) \subset \neg\neg s(B) \subset \neg\neg U \subset \neg - U.$$

Choosing N such that $x_n \in -U$ for all $n \succcurlyeq N$, we now see that

$$B \subset \neg \{n : n \succcurlyeq N\}.$$

Now suppose, conversely, that (2.17) holds. If $x \in -U$, we apply the weak nested neighbourhoods property, to obtain $V \subset X$ such that $x \in -V$ and $\neg V \subset -U$. With

$$B \equiv \{n \in \mathfrak{D} : x_n \in V\},$$

we see that $x \bowtie s(B)$; so there exists $N \in \mathfrak{D}$ such that

$$B \subset \neg \{n : n \succcurlyeq N\}.$$

If $n \succcurlyeq N$, then $x_n \in \neg V \subset -U$. ∎

A topological space (X, τ) has an associated notion of topological convergence: a net $(x_n)_{n \in \mathfrak{D}}$ in X is said to **converge topologically** to $x \in X$ if for each neighbourhood U of x in X there exists $N \in \mathfrak{D}$ such that $x_n \in U$ for all $n \succcurlyeq N$. Since the $\bowtie_\tau$-nearly open subsets of X are τ-open, we see that if a net in X converges topologically to a limit $x \in X$, then it converges to x relative to the topological pre-apartness on X.

Proposition 2.4.3 *The following are equivalent conditions on a topological pre-apartness space (X, τ):*

(i) *Apartness convergence is equivalent to topological convergence.*

(ii) *Apartness convergence implies topological convergence.*

(iii) *X is topologically consistent.*

Proof. In view of the remark immediately preceding this proposition, it is clear that (i) and (ii) are equivalent.

Supposing that (ii) holds, consider a τ-open set U and any $x \in U$. Let $(x_n)_{n \in \mathfrak{D}_x}$ be the basic neighbourhood net of x, which converges to x relative to the apartness $\bowtie$ associated with τ. By (ii), this net converges topologically to x, so there exists $N \equiv (\xi, V) \in \mathfrak{D}_x$ such that $x_n \in U$ for all $n \succcurlyeq N$.

For each $y \in -V$ we have $(y, V) \succcurlyeq N$, by definition of the reverse inclusion preorder; so $y = x_{(y,V)} \in U$. Hence $x \in -V \subset U$. It follows that U is a union of apartness complements and is therefore nearly open. Hence (ii) implies (iii).

The proof that (iii) implies (ii) is left as an exercise. ∎

Recall that when we refer to the closure $\overline{A}$ of a subset A of a pre-apartness space X, we mean the closure of A with respect to the apartness topology on X.

Proposition 2.4.4 *The closure of a subset A of a pre-apartness space X consists of all points of X that are limits of nets in A.*

Proof. If $(x_n)_{n \in \mathfrak{D}}$ is a net in A converging to an element x of X, then, clearly,

$$\mathrm{near}\,(x, \{x_n : n \in \mathfrak{D}\})$$

and therefore near (x, A); whence $x \in \overline{A}$, by Proposition 2.2.17.

Conversely, given x in $\overline{A}$, let

$$\mathfrak{D} \equiv \{(y, U) : x \in -U \wedge y \in A - U\}\,.$$

Then $\mathfrak{D}$ is directed by the usual reverse inclusion preorder $\succcurlyeq$. Let $(y_n)_{n \in \mathfrak{D}}$ be the net in A defined by the mapping $(y, U) \rightsquigarrow y$, and let $U \subset X$ be such that $x \in -U$. Since near (x, A) (by Proposition 2.2.17), there exists $y \in A - U$; let $n_0 \equiv (y, U) \in \mathfrak{D}$. For each $n \equiv (y_n, V)$ in $\mathfrak{D}$ with $n \succcurlyeq n_0$ we have $x \in -V$ and $y_n \in A - V \subset -V \subset -U$. Thus $(y_n)_{n \in \mathfrak{D}}$ converges to x. ∎

Since it is possible for a net in a pre-apartness space to have more than one limit, it is reasonable to look for a characterisation of those pre-apartness spaces in which a convergent net has a unique limit. With an eye on classical topology, we say that a topological space (X, τ), or just its topology τ, is **(topologically) Hausdorff** if the following condition holds:

> For all x, y in X with $x \neq y$, there exist U, V in τ such that $x \in U, y \in V$, and $U \subset \sim V$.

On the other hand, we say that a pre-apartness space $(X, \bowtie)$, or just its pre-apartness $\bowtie$, is **(apartness) Hausdorff** if it satisfies the following condition:

> For all $x, y \in X$ with $x \neq y$, there exist $U, V \subset X$ such that $x \in -U, y \in -V$, and $-U \subset \sim -V$.

In that case, $-V \subset \sim -U$.

Note that for an *apartness* to be Hausdorff, it suffices that the following hold:

> If $x, y \in X$ and $x \neq y$, then there exist $U, V \subset X$ such that $x \in -U, y \in -V$, and $-U \cap -V = \varnothing$.

For if we have such U and V, then for all $s \in -U$ and $t \in -V$, either $s \neq t$ or $s \in -V$, by **A5**; but the latter is ruled out, since $-U \cap -V = \varnothing$. Hence $-U \subset \sim - V$.

It should be clear that a pre-apartness is Hausdorff if and only if the corresponding apartness topology is Hausdorff.

Proposition 2.4.5 *Let (X, τ) be a topological space. If X is topologically Hausdorff, then $\bowtie_\tau$ is Hausdorff. Conversely, if $\bowtie_\tau$ is Hausdorff and X is topologically consistent, then X is topologically Hausdorff.*

Proof. Suppose first that X is topologically Hausdorff, and consider points $x, y \in X$ with $x \neq y$. Pick open subsets A, B of X with $x \in A, y \in B$, and $A \subset \sim B$. Since $x \in A \subset \sim \sim A$, we have $x \in -_\tau \sim A$, where $-_\tau$ denotes the apartness complement corresponding to $\bowtie_\tau$. Similarly, $y \in -_\tau \sim B$. Moreover,

$$-_\tau \sim B \subset -_\tau A \subset \sim A,$$

so

$$-_\tau \sim A \subset -_\tau -_\tau \sim B \subset \sim -_\tau \sim B.$$

Hence $\bowtie_\tau$ is Hausdorff.

Now suppose, conversely, that $\bowtie_\tau$ is Hausdorff. The remark preceding this proposition shows that the apartness topology corresponding to $\bowtie_\tau$ is Hausdorff. It follows that if (X, τ) is topologically consistent, then it is topologically Hausdorff. ∎

Proposition 2.4.6 *A Hausdorff pre-apartness space is $\mathbf{T}_1$.*

Proof. Let X be a Hausdorff pre-apartness space, and let x, y be points of X with $x \neq y$. There exist $U, V \subset X$ such that $x \in -U \subset \sim - V$ and $y \in -V$. Applying **A4**, we see that $-U \subset -- V$, so $x \bowtie -V$. Since $\{y\} \subset -V$, it follows that $x \bowtie \{y\}$. ∎

Classically, being Hausdorff is equivalent to having the **unique limits property**,

ULP: *If $(x_n)_{n \in \mathcal{D}}$ is a net converging to limits x and y in X, then $x = y$.*

From a constructive viewpoint, the unique limits property appears rather weak. To introduce a stronger constructive property, we say that a point y in X is **eventually bounded away from a net** $(x_n)_{n \in \mathcal{D}}$ in X if there exists $n_0 \in \mathcal{D}$ such that

$$y \in - \{x_n : n \succcurlyeq n_0\}.$$

We now state the **strong unique limits property** (classically equivalent to the unique limits property),

SULP: *If $(x_n)_{n \in \mathcal{D}}$ is a net in X that converges to a limit x, and if $x \neq y$ in X, then $(x_n)_{n \in \mathcal{D}}$ is eventually bounded away from y.*

It is straightforward to show that in the presence of a tight inequality, **SULP** implies **ULP**.

Not surprisingly (in view of classical topology), the Hausdorff condition is linked to our two types of uniqueness of limits.

Proposition 2.4.7 *A pre-apartness space is Hausdorff if and only if it has the strong unique limits property.*

Proof. Let X be a pre-apartness space. Assume first that X is Hausdorff, let $(x_n)_{n \in \mathfrak{D}}$ be a net converging to a limit x in X, and let $x \neq y$ in X. Choose U,V such that $x \in -U, y \in -V$, and $-U \subset \sim -V$. There exists n_0 such that $x_n \in -U$ for all $n \succcurlyeq n_0$. Then

$$y \in -V \subset \sim - U \subset \sim \{x_n : n \succcurlyeq n_0\},$$

so, by **A4**,

$$y \in -\{x_n : n \succcurlyeq n_0\}.$$

Hence **SULP** holds in X.

Now suppose, conversely, that X has the strong unique limits property. Let x,y be points of X with $x \neq y$. Since the net $\mathcal{N}_x$ converges to x, **SULP** ensures that there exists $n_0 \equiv (\xi, U) \in \mathfrak{D}_x$ such that

$$y \in -\{x_n : n \in \mathfrak{D}_x, \ n \succcurlyeq n_0\}.$$

By the definition of $\mathfrak{D}_x$ and Lemma 2.4.1,

$$x \in -U = \{x_n : n \in \mathfrak{D}_x, \ n \succcurlyeq n_0\}.$$

It follows that $y \in --U \subset \sim -U$. Hence X is Hausdorff. ∎

From this proposition and a remark just before it, we obtain

Corollary 2.4.8 *A Hausdorff pre-apartness space with tight inequality has the unique limits property.*

Corollary 2.4.8 has a noteworthy converse.

Proposition 2.4.9 *In an apartness space with the unique limits property the inequality is tight.*

Proof. Let X be such an apartness space, and let x,y be points of X with $\neg(x \neq y)$. Using this last property and **A5**, we see that

$$\forall_{U \subset X}(x \in -U \Leftrightarrow y \in -U).$$

It readily follows that the net $\mathcal{N}_x$ converges to both x and y; whence, by **ULP**, $x = y$. ∎

It is worth taking a little time out of the main development to show that the connections established in the preceding three results are the best possible within our constructive framework. We begin by showing that Hausdorff/**SULP** is not enough to establish tightness.

Proposition 2.4.10 *If every apartness space that has the strong unique limits property (or, equivalently, is Hausdorff) has tight inequality, then the law of excluded middle holds.*

Proof. Let P be any statement such that $\neg\neg P$ holds, and take $X \equiv \{0,1,2\}$ with equality satisfying

$$0 = 1 \Leftrightarrow P$$

and inequality given by

$$0 \neq 2, \ 1 \neq 2, \ \text{and} \ (0 \neq 1 \Leftrightarrow \neg P).$$

Define a point-set pre-apartness $\bowtie$ on X by

$$x \bowtie A \Leftrightarrow x \in \sim A.$$

We show that $\bowtie$ is, in fact, an apartness. Consider all possible cases that arise when $x \in -A$. If $x = 0$, then $0 \in \sim A$. It follows that $A \subset \{2\}$: for if $y \in A$, then either $y = 1$ or $y = 2$; in the former case, $0 \neq 1$ (since $0 \in \sim A$) and therefore $\neg P$ holds, which contradicts our hypotheses. Since $1 \neq 2$, we have $1 \in \sim A$; since also $0 \neq 2$, we conclude that

$$\forall_{y \in X} (0 \neq y \vee y \in \sim A = -A).$$

The case $x = 1$ is similar, and the case $x = 2$ is even easier to handle.

We claim that X is a Hausdorff apartness space and hence, by Proposition 2.4.7, has the strong unique limits property. If $x \neq y$, then without loss of generality, either $x = 0$ and $y = 2$, or else $x = 1$ and $y = 2$. Taking, for illustration, the former case, we have

$$0 \in \{0,1\} = \sim \{2\} = -\{2\},$$

$$2 \in \{2\} = \sim \{0\} = -\{0\},$$

and

$$\sim\{2\} = \sim\sim\{0\}.$$

Thus there exist $U \equiv \{2\}$ and $V \equiv \{0\}$ such that $0 \in -U$, $2 \in -V$, and $-U \subset \sim -V$.

Finally, if the inequality on X is tight, then as $\neg (0 \neq 1)$, we have $0 = 1$ and therefore P. ∎

Next we aim to show that even for a locally decomposable[3] apartness space, the unique limits property does not entail being Hausdorff. For the proof we introduce a strange lemma and a general construction. The lemma may seem obvious, but we need it in order to avoid the axiom of choice.

---

[3] The reader should by now have realised that local decomposability is more than handy for tidying up loose ends in cases like this; but one should beware of putting too much trust in it, since there are situations for which even local decomposability is not strong enough to enable us to recover constructively the full form of a classical theorem about apartness.

Lemma 2.4.11 *Let $\mathcal{C}$ be a class of subsets of a set X, and let $(S_i)_{i \in I}$ be a family of subsets of X such that for each i, if S_i is inhabited, then it is a union of sets in $\mathcal{C}$. If $S \equiv \bigcup_{i \in I} S_i$ is inhabited, then it is also a union of sets in $\mathcal{C}$.*

Proof. It is straightforward to verify that

$$S = \bigcup \{U \in \mathcal{C} : \exists_{x \in X} \exists_{i \in I} (x \in U \subset S_i)\},$$

which gives exactly what we want. ∎

Let X be a set with an inequality $\neq$ that is **cotransitive**, in the sense that

$$\forall_{x,y \in X} (x \neq y \Rightarrow \forall_{z \in X} (x \neq z \vee z \neq y)).$$

We say that a subset S of X is **cofinite** if it is the complement of a finitely enumerable subset. If X has at least two distinct points, then we define the **cofinite topology** on X to be

$$\tau_{\text{cof}} \equiv \{S \subset X : S \neq \varnothing \Rightarrow S \text{ is a union of cofinite sets}\}.$$

To see that this is a topology, first pick distinct points x,y of X. By cotransitivity,

$$X = \sim\{x\} \cup \sim\{y\},$$

so $X \in \tau_{\text{cof}}$. Also, by *ex falso quodlibet*, $\varnothing \in \tau_{\text{cof}}$. The unions axiom for a topology is an immediate consequence of Lemma 2.4.11 with $\mathcal{C} \equiv \tau_{\text{cof}}$. To verify the intersections axiom, let $(A_i)_{i \in I}$ and $(B_j)_{j \in J}$ be families of finitely enumerable subsets of X, and consider the elements

$$S \equiv \bigcup_{i \in I} \sim A_i, \ T \equiv \bigcup_{j \in J} \sim B_j$$

of τ_{cof}. We have

$$S \cap T = \left(\bigcup_{i \in I} \sim A_i\right) \cap \left(\bigcup_{j \in J} \sim B_j\right)$$
$$= \bigcup_{i \in I} \bigcup_{j \in J} (\sim A_i \cap \sim B_j)$$
$$= \bigcup_{i \in I} \bigcup_{j \in J} \sim (A_i \cup B_j),$$

where each of the sets $A_i \cup B_j$ is finitely enumerable. Hence $S \cap T \in \tau_{\text{cof}}$.

Proposition 2.4.12 *If every locally decomposable, $\mathbf{T}_1$ apartness space that contains two distinct points and has the unique limits property is Hausdorff, then Markov's principle holds.*

Proof. We take a specific case of the foregoing construction. Let $(a_n)_{n \geqslant 1}$ be a decreasing binary sequence such that

$$a_1 = 1 \wedge \neg \forall_n (a_n = 1).$$

Take
$$X \equiv \{0\} \cup \left\{ \frac{a_n}{n} : n = 1, 2, 3, \dots \right\}$$

with the discrete equality and inequality, let τ be the cofinite topology on X, and let $\bowtie$ be the corresponding pre-apartness. To show that X is topologically locally decomposable and hence (by Corollary 2.2.16 and Proposition 2.2.13) an apartness space, consider $x \in X$ and $U \in \tau$ with $x \in U$. Since X contains two distinct points, we may assume that $U = \sim A$ for some finitely enumerable (and hence, in this case, finite) set $A \subset X$; without loss of generality we may assume that A is inhabited. Consider first the case $x = 0$, in which

$$\varnothing \neq A \subset \left\{ \frac{1}{n} : n = 1, 2, 3, \dots \right\}.$$

Let
$$K \equiv \max \left\{ k : \frac{1}{k} \in A \right\} \tag{2.18}$$

and
$$V \equiv \{0\} \cup \left\{ \frac{a_k}{k} : k > K \right\} = \sim \left\{ \frac{a_k}{k} : k \leqslant K \right\}.$$

Then V is a neighbourhood of 0. For each $y \in X$, either $y = 0 \in U$ or else $y = 1/k$ for some k with $a_k = 1$. In the latter case, if $k > K$, then $y \in \sim A = U$; whereas if $k \leqslant K$, then $y \in \sim V$. This deals with the case $x = 0$. Now consider the case where $x = 1/m$ for some m with $a_m = 1$. If $a_{m+1} = 0$, then X is finite and hence topologically locally decomposable; so we may assume that $a_{m+1} = 1$. Since $x \neq 1/(m+1)$, we may further assume that $1/(m+1) \in A$. Thus $K > m$, where K is defined as at (2.18). Set

$$W \equiv \left\{ \frac{a_k}{k} : (k > K \wedge a_k = 1) \vee k = m \right\}$$
$$= \sim \left(\{0\} \cup \left\{ \frac{a_k}{k} : m \neq k \leqslant K \right\} \right).$$

Then W is a neighbourhood of x. For each $y \in X$, either $y = 0$ and hence $y \in \sim W$, or else $y = a_k/k$ for some k with $a_k = 1$. If $k > K$, then $y \in \sim A = U$; if $k = m$, then $y = x \in U$; if $m \neq k \leqslant K$, then $y \in \sim W$. Thus, taking all the various cases together, we see that the pre-apartness space X is topologically locally decomposable.

We now confirm that the cofinite topology is $\mathbf{T}_1$. To this end, let $x \neq y$ in X. Either one of the points x and y is 0 or else both are nonzero. If, for example, $x = 0$, then $y = 1/n$ for some n with $a_n = 1$. Writing

$$U \equiv \{0\} \cup \left\{ \frac{a_k}{k} : k > n \right\} = \sim \left\{ \frac{a_k}{k} : k \leqslant n \right\},$$

we see that
$$U \in \tau \text{ and } x \in U \subset \sim \{y\}. \tag{2.19}$$

So we are left with the case where $x = 1/m$ and $y = 1/n$, with $a_m = a_n = 1$. In this case, without loss of generality taking $m > n$, we obtain (2.19) by defining

$$U \equiv \{0\} \cup \left\{ \frac{a_k}{k} : k \geqslant m \right\}.$$

Next, we prove that X has the unique limits property. Suppose that $(x_n)_{n \in \mathcal{D}}$ is a net in X that converges to both x and y. Suppose also that $x \neq y$. For each n, if $a_n = 0$, then X is finite and so has the unique limits property; whence $x = y$, a contradiction. Thus $a_n = 1$ for all n, which is also a contradiction. We conclude that $\neg (x \neq y)$; since we are dealing with a discrete inequality, it follows that $x = y$.

Finally, noting that $0 \neq 1$, suppose there exist $U, V \subset X$ such that $0 \in -U, 1 \in -V$, and $-U \cap -V = \varnothing$. Pick finitely enumerable sets $A, B \subset X$ such that $0 \in {\sim} A \subset -U$ and $1 \in {\sim} B \subset -V$. Let

$$N \equiv \max \left\{ n : \frac{1}{n} \in A \cup B \right\}.$$

If $a_{N+1} = 1$, then

$$\frac{1}{N+1} \in {\sim} A \cap {\sim} B \subset -U \cap -V,$$

a contradiction; hence $a_{N+1} = 0$. Recalling from page 43 the remark about the Hausdorff property in an *apartness* space, we see that if X is Hausdorff, then there exists N such that $a_n = 0$ for all $n > N$. ∎

Having dealt with the convergence of nets, we now turn to the convergence of functions. Let X, Y be pre-apartness spaces, x a point of X such that near $(x, {\sim} \{x\})$, and f a mapping of ${\sim} \{x\}$ into Y. We say that $y \in Y$ is a **limit of f at x**, or a **limit of $f(t)$ as t tends to x**, if the net $f(\mathcal{N}'_x)$ converges to y in Y. We then write

$$f(t) \rightarrow y \text{ as } t \rightarrow x.$$

In the case where the limit of $f(\mathcal{N}'_x)$ is unique, we also write

$$\lim_{t \rightarrow x, \ t \in X} f(t) = y$$

or just

$$\lim_{t \rightarrow x} f(t) = y.$$

Proposition 2.4.13 *Let X and Y be pre-apartness spaces, x a point of X such that near $(x, {\sim} \{x\})$, y a point of Y, and f a mapping of ${\sim} \{x\}$ into Y. Then the following conditions are equivalent:*

(i) *y is a limit of f at x.*

(ii) $\forall_{V \subset Y} (y \in -V \Rightarrow \exists_{U \subset X} (x \in -U \wedge f(-U \cap \sim \{x\}) \subset -V))$.

Proof. Let $y \in -V \subset Y$. Suppose that y is a limit of f at x; then there exists $n_0 \in \mathfrak{D}'_x$ such that $f(x_n) \in -V$ for all $n \succcurlyeq n_0$. Writing $n_0 \equiv (\xi, U)$, we have $x \in -U$; moreover, by Lemma 2.4.1,

$$\{x_n : n \succcurlyeq n_0\} = -U \cap \sim \{x\}.$$

It follows that

$$f(-U \cap \sim \{x\}) \subset -V. \tag{2.20}$$

Thus (i) implies (ii).

Conversely, assume (ii). With y and V as before, choose $U \subset X$ such that $x \in -U$ and (2.20) holds. Since near $(x, \sim \{x\})$, there exists $\xi \in -U \cap \sim \{x\}$. Let $\nu \equiv (\xi, U) \in \mathfrak{D}'_x$. By Lemma 2.4.1,

$$-U \cap \sim \{x\} = \{x_n : n \in \mathfrak{D}'_x, n \succcurlyeq \nu\}.$$

It follows from (2.20) that $f(x_n) \in -V$ for all $n \succcurlyeq \nu$. Hence y is a limit of f at x. ∎

Proposition 2.4.14 *Let X,Y,Z be pre-apartness spaces such that Z has the weak nested neighbourhoods property. Let $x \in X$ and $y \in Y$, let f be a mapping of $\sim \{x\}$ into Y, and let g be a mapping of Y into Z. Suppose that y is a limit of f at x, and that g is continuous at y. Then $g(y)$ is a limit of $g \circ f$ at x.*

Proof. Let $g(y) \in -W \subset Z$. Using **WNN**, choose $A \subset Z$ such that $g(y) \in -A$ and $\neg A \subset -W$. Then, by continuity, $y \in -g^{-1}(A) \subset Y$, so, by Proposition 2.4.13, there exists $U \subset X$ such that $x \in -U$ and $f(-U \cap \sim \{x\}) \subset -g^{-1}(A)$. Since near $(x, \sim \{x\})$, there exists $\xi \in -U$ such that $\xi \neq x$. Then $n_0 \equiv (\xi, U)$ belongs to $\mathfrak{D}'_x$; also, by Lemma 2.4.1, for each $n \succcurlyeq n_0$ in $\mathfrak{D}'_x$ we have $x_n \in -U \cap \sim \{x\}$ and therefore $f(x_n) \in -g^{-1}(A)$; whence $(g \circ f)(x_n) \notin A$ and therefore $(g \circ f)(x_n) \in -W$. ∎

Proposition 2.4.15 *Let X,Y be pre-apartness spaces, $x \in X$, $y \in Y$, and f a mapping of $\sim \{x\}$ into Y such that y is a limit of f at x. Then f preserves the convergence of nets at x : that is, for each net $(x_n)_{n \in \mathfrak{D}}$ in $\sim \{x\}$ that converges to x in X, the net $(f(x_n))_{n \in \mathfrak{D}}$ converges to y.*

Proof. Let $(x_n)_{n \in \mathfrak{D}}$ be a net in $\sim \{x\}$ that converges to x in X, and let $y \in -V \subset Y$. By Proposition 2.4.13, there exists $U \subset X$ such that $x \in -U$ and $f(-U \cap \sim \{x\}) \subset -V$. Choose n_0 in $\mathfrak{D}$ such that $x_n \in -U$ for all $n \succcurlyeq n_0$ in $\mathfrak{D}$. For such n we have $x_n \in -U \cap \sim \{x\}$ and therefore $f(x_n) \in -V$. ∎

Proposition 2.4.16 *Let X,Y be pre-apartness spaces, f a topologically continuous mapping of X into Y, and $(x_n)_{n \in \mathfrak{D}}$ a net that converges to a limit x in X. Then the net $(f(x_n))_{n \in \mathfrak{D}}$ converges in Y to $f(x)$.*

Proof. Let $f(x) \in -V$ in Y; then $f^{-1}(-V)$ is nearly open in V. Pick $U \subset X$ such that $x \in -U \subset f^{-1}(-V)$. There exists $n_0 \in \mathfrak{D}$ such that $x_n \in -U$, and therefore $f(x_n) \in -V$, for all $n \succcurlyeq n_0$. ∎

Corollary 2.4.17 *Let X be an apartness space, Y a pre-apartness space with the weak nested neighbourhoods property, $f : X \to Y$ a continuous mapping, and $(x_n)_{n \in \mathfrak{D}}$ a net that converges to a limit x in X. Then the net $(f(x_n))_{n \in \mathfrak{D}}$ converges in Y to $f(x)$.*

Proof. Apply Corollary 2.3.8 and Proposition 2.4.16. ∎

Limits will re-appear in Chapter 3, in the context of complete pre-apartness spaces.

2.5   Product Pre-apartness Spaces

Let X_1 and X_2 be pre-apartness spaces, let $X \equiv X_1 \times X_2$, and recall our convention that, for example, $\mathbf{x}$ denotes the element (x_1, x_2) of X. We define the relation $\bowtie$ between points and subsets of X as follows:

$$\mathbf{x} \bowtie A \Leftrightarrow \exists_{U_1 \subset X_1} \exists_{U_2 \subset X_2} (\mathbf{x} \in -U_1 \times -U_2 \subset {\sim}A), \qquad (2.21)$$

where for $k = 1, 2$ the set $-U_k$ is the apartness complement of U_k in the pre-apartness space X_k.

Proposition 2.5.1 *The relation $\bowtie$ defined at (2.21) is a pre-apartness on $X \equiv X_1 \times X_2$.*

Proof. For each $\mathbf{x} \in X$, since

$$\mathbf{x} \in X_1 \times X_2 = -\emptyset \times -\emptyset = {\sim}\emptyset,$$

we see that $\mathbf{x} \bowtie \emptyset$. It is clear from (2.21) that $-A \subset {\sim}A$ in $(X, \bowtie)$. If $\mathbf{x} \bowtie A \cup B$ in X, then there exist $U_k \subset X_k$ such that

$$\mathbf{x} \in -U_1 \times -U_2 \subset {\sim}(A \cup B) = {\sim}A \cap {\sim}B,$$

so $\mathbf{x} \bowtie A$ and $\mathbf{x} \bowtie B$. If, conversely, $\mathbf{x} \bowtie A$ and $\mathbf{x} \bowtie B$, then there exist $U_k, V_k \subset X_k$ such that

$$\mathbf{x} \in -U_1 \times -U_2 \subset {\sim}A,$$
$$\mathbf{x} \in -V_1 \times -V_2 \subset {\sim}B,$$

and therefore

$$\mathbf{x} \in (-U_1 \cap -V_1) \times (-U_2 \cap -V_2) \subset {\sim}A \cap {\sim}B.$$

Referring to **A3** in the space X_k, we now see that

$$\mathbf{x} \in -(U_1 \cup V_1) \times -(U_2 \cup V_2) \subset {\sim}(A \cup B).$$

Hence $\mathbf{x} \bowtie A \cup B$. This completes the verification of **A3** in X.

Finally, if $-U_1 \times -U_2 \subset {\sim}A$, then the definition (2.21) shows that $-U_1 \times -U_2 \subset -A$; so if also $-A \subset {\sim}B$, then $-U_1 \times -U_2 \subset {\sim}B$ and therefore $-U_1 \times -U_2 \subset -B$. It follows from this that **A4** holds in X. ∎

We call the pre-apartness defined at (2.21) the **product pre-apartness** on X. Equipped with the usual inequality (see page 10) and the foregoing pre-apartness structure, X is known as the **product of the pre-apartness spaces** X_1 and X_2. The corresponding nearness on X is then given by

$$\mathrm{near}(\mathbf{x}, A) \Leftrightarrow \forall_{B \subset X} \left(\mathbf{x} \bowtie B \Rightarrow \exists_{\mathbf{y} \in A} \left(\mathbf{y} \bowtie B \right) \right).$$

Our next aim is to show that some of the most important properties hold in the product pre-apartness space if and only if they hold in each of its factors. We first prove two key lemmas, for which we note that (as is easily demonstrated) if $A_k \subset X_k$, then

$$\sim A_1 \times X_2 = {\sim}(A_1 \times X_2)$$

and

$$X_1 \times {\sim}A_2 = {\sim}(X_1 \times A_2).$$

Lemma 2.5.2 *Let $X \equiv X_1 \times X_2$ be the product of two pre-apartness spaces, and let $A_k \subset X_k$ $(k = 1, 2)$. Then $-A_1 \times X_2 = -(A_1 \times X_2)$ and $X_1 \times -A_2 = -(X_1 \times A_2)$.*

Proof. Since $-A_1 \times X_2 = -A_1 \times -\varnothing$ and

$$-A_1 \times X_2 \subset {\sim}A_1 \times X_2 = {\sim}(A_1 \times X_2),$$

the definition of the product apartness shows that $-A_1 \times X_2 \subset -(A_1 \times X_2)$. Conversely, given (x_1, x_2) in $-(A_1 \times X_2)$, we can find subsets U_k of X_k such that

$$(x_1, x_2) \in -U_1 \times -U_2 \subset {\sim}(A_1 \times X_2) = {\sim}A_1 \times X_2$$

and therefore $x_1 \in -U_1 \subset {\sim}A_1$. It now follows from **A4** in the space X_1 that $x_1 \bowtie A_1$. Thus $-(A_1 \times X_2) \subset -A_1 \times X_2$. The other part of the lemma is proved similarly. ∎

Lemma 2.5.3 *If X_k is a pre-apartness space and $U_k \subset X_k$ $(k = 1, 2)$, then*

$$-U_1 \times -U_2 = -\left((U_1 \times X_2) \cup (X_1 \times U_2) \right).$$

Proof. By Lemma 2.5.2 and **A3**,

$$
\begin{aligned}
-U_1 \times -U_2 &= (-U_1 \times X_2) \cap (X_1 \times -U_2) \\
&= -(U_1 \times X_2) \cap -(X_1 \times U_2) \\
&= -((U_1 \times X_2) \cup (X_1 \times U_2)),
\end{aligned}
$$

as we required. ∎

Proposition 2.5.4 *The product pre-apartness space $X \equiv X_1 \times X_2$ satisfies* **A5** *if and only if both X_1 and X_2 satisfy* **A5**.

Proof. Suppose first that X satisfies **A5**, fix $x_2 \in X_2$, and let $x_1 \bowtie A_1$ in X_1. Then, by Lemma 2.5.2,

$$
(x_1, x_2) \in -A_1 \times X_2 = -(A_1 \times X_2),
$$

so $(x_1, x_2) \bowtie A_1 \times X_2$ in X. Applying **A5** in X, for each $x \in X_1$ we have either $(x_1, x_2) \neq (x, x_2)$ and therefore $x_1 \neq x$, or else $(x, x_2) \bowtie A_1 \times X_2$. In the latter case, by Lemma 2.5.2, $(x, x_2) \in -A_1 \times X_2$; so $x \in -A_1$, and therefore $x \bowtie A_1$, in X_1. This completes the proof of **A5** in the pre-apartness space X_1. The proof for X_2 is similar.

Now suppose, conversely, that **A5** holds in both X_1 and X_2. Let $\mathbf{x} \bowtie A$ in $X_1 \times X_2$, and choose sets $U_k \subset X_k$ such that $\mathbf{x} \in -U_1 \times -U_2 \subset {\sim}A$; then $x_k \in -U_k$. Consider any $\mathbf{y} \in X$. Applying **A5** in the pre-apartness space X_1, we have either $x_1 \neq y_1$ or $y_1 \in -U_1$. Since $\mathbf{x} \neq \mathbf{y}$ in the first case, we may assume that $y_1 \in -U_1$. Likewise, we may assume that $y_2 \in -U_2$. Hence $\mathbf{y} \in -U_1 \times -U_2 \subset {\sim}A$ and therefore $\mathbf{y} \bowtie A$. This completes the verification of **A5** in the space X. ∎

Proposition 2.5.5 *The product pre-apartness space $X \equiv X_1 \times X_2$ is* $\mathbf{T}_1$ *if and only if both X_1 and X_2 are* $\mathbf{T}_1$.

Proof. Suppose that X is $\mathbf{T}_1$, fix $x_2 \in X_2$, and let $x \neq y$ in X_1. Then $(x, x_2) \neq (y, x_2)$ in X; so, by the $\mathbf{T}_1$ property in X, $(x, x_2) \bowtie \{(y, x_2)\}$ and there exist $U_k \subset X_k$ such that

$$
(x, x_2) \in -U_1 \times -U_2 \subset {\sim} \{(y, x_2)\}.
$$

It follows that $x \in -U_1$ and $-U_1 \times \{x_2\} \subset {\sim}\{(y, x_2)\}$; thus $x \in -U_1 \subset {\sim}\{y\}$, and therefore $x \bowtie \{y\}$, by **A4** in X_1. Hence X_1, and similarly X_2, is a $\mathbf{T}_1$ pre-apartness space.

Now suppose, conversely, that both X_1 and X_2 are $\mathbf{T}_1$ pre-apartness spaces. Let $\mathbf{x} \neq \mathbf{y}$ in X. Then either $x_1 \neq y_1$ or else $x_2 \neq y_2$. Taking, for example, the first alternative and applying the $\mathbf{T}_1$ property in X_1, we see that $x_1 \bowtie \{y_1\}$; whence $\mathbf{x} \in -\{y_1\} \times X_2 \subset {\sim}\{\mathbf{y}\}$. It follows that $\mathbf{x} \bowtie \{\mathbf{y}\}$. Hence the $\mathbf{T}_1$ property holds for X. ∎

Proposition 2.5.6 *The product $X \equiv X_1 \times X_2$ of two pre-apartness spaces is locally decomposable if and only if both X_1 and X_2 are locally decomposable.*

Proof. Suppose first that X is locally decomposable. In order to prove that X_1 is locally decomposable, let $x_1 \in -U_1 \subset X_1$, and pick $x_2 \in X_2$. Then (note Lemma 2.5.2)

$$\mathbf{x} \equiv (x_1, x_2) \in -U_1 \times X_2 = -(U_1 \times X_2).$$

Hence, by the local decomposability of X, there exists $T \subset X$ such that

$$\mathbf{x} \in -T \wedge \forall_{\mathbf{y} \in X} \left(\mathbf{y} \in -(U_1 \times X_2) \vee \mathbf{y} \in T \right).$$

Let

$$V_1 \equiv \{ \xi \in X_1 : (\xi, x_2) \in T \}.$$

It will suffice to show that $x_1 \in -V_1$ and $X_1 = -U_1 \cup V_1$. Since $\mathbf{x} \in -T$, we can find $W_k \subset X_k$ such that $\mathbf{x} \in -W_1 \times -W_2 \subset \sim T$. For any $x \in -W_1$ and $v \in V_1$ we have $(x, x_2) \in -W_1 \times -W_2 \subset \sim T$ and $(v, x_2) \in T$; whence $x \neq v$. Thus $-W_1 \subset \sim V_1$. Since $x_1 \in -W_1$, it follows from **A4** in X_1 that $x_1 \in -V_1$, as required. On the other hand, given x in X_1, we have either $(x, x_2) \in -(U_1 \times X_2)$ and therefore (by Lemma 2.5.2) $x \in -U_1$, or else $(x, x_2) \in T$ and so $x \in V_1$. Thus X_1, and similarly X_2, is locally decomposable.

Now suppose, conversely, that each X_k is locally decomposable. Consider $\mathbf{x} \in X$ and $S \subset X$ such that $\mathbf{x} \in -S$. Choose $U_1 \subset X_1$ and $U_2 \subset X_2$ such that $\mathbf{x} \in -U_1 \times -U_2 \subset \sim S$. Since $x_k \in -U_k$, there exists $V_k \subset X_k$ such that

$$x_k \in -V_k \wedge \forall_{x \in X_k} (x \in -U_k \vee x \in V_k).$$

Let

$$T \equiv (V_1 \times X_2) \cup (X_1 \times V_2).$$

Then $\mathbf{x} \in -V_1 \times -V_2 \subset -T$, by Lemma 2.5.3. On the other hand, for each $\xi \in X$, either $\xi_1 \in -U_1$ and $\xi_2 \in -U_2$, in which case $\xi \in -S$; or else we have either $\xi_1 \in V_1$ or $\xi_2 \in V_2$, and so $\xi \in T$; thus $X = -S \cup T$. Hence X is locally decomposable. ∎

Proposition 2.5.7 *The product $X \equiv X_1 \times X_2$ of two pre-apartness spaces is Hausdorff if and only if both X_1 and X_2 are Hausdorff.*

Proof. Suppose that X is Hausdorff, fix $x_2 \in X_2$, and let $x \neq y$ in X_1. Then $(x, x_2) \neq (y, x_2)$ in X; so there exist $U, V \subset X$ such that $(x, x_2) \in -U, (y, x_2) \in -V$, and $-U \subset \sim -V$. By definition of the product pre-apartness, there exist $U_k \subset X_k$ and $V_k \subset X_k$ such that $(x, x_2) \in -U_1 \times -U_2 \subset \sim U$ and $(y, x_2) \in -V_1 \times -V_2 \subset \sim V$. Then $x \in -U_1$ and $y \in -V_1$. Moreover,

$$-U_1 \times \{x_2\} \subset -U_1 \times -U_2 \subset -U \subset \sim -V \subset \sim(-V_1 \times \{x_2\}),$$

so $-U_1 \subset \sim -V_1$. Hence X_1, and similarly X_2, is Hausdorff.

Now suppose, conversely, that both X_1 and X_2 are Hausdorff. Let $\mathbf{x} \neq \mathbf{y}$ in X. Then either $x_1 \neq y_1$ or else $x_2 \neq y_2$. Taking, for example, the alternative $x_1 \neq y_1$ and applying the Hausdorff property in X_1, we obtain $U_1, V_1 \subset X_1$ such that $x_1 \in -U_1, y_1 \in -V_1$, and $-U_1 \subset \sim -V_1$. Then, by Lemma 2.5.2 and the comment preceding it,

$$\mathbf{x} \in -U_1 \times X_2 \subset (\sim -V_1) \times X_2$$
$$= \sim(-V_1 \times X_2) = \sim - (V_1 \times X_2).$$

Since (again by Lemma 2.5.2) $\mathbf{y} \in -(V_1 \times X_2)$, we conclude that X is Hausdorff. ∎

Classically, for the product proximity structure on X we have near$(\mathbf{x}, A)$ if and only if $\neg (\mathbf{x} \bowtie A)$, which is equivalent to the condition

$$\forall_{U_1 \subset X_1} \forall_{U_2 \subset X_2} (\mathbf{x} \in -U_1 \times -U_2 \Rightarrow \exists_{\mathbf{y}} (\mathbf{y} \in (-U_1 \times -U_2) \cap A)). \quad (2.22)$$

Constructively, we have

Proposition 2.5.8 *Let $X \equiv X_1 \times X_2$ be a product of two pre-apartness spaces, let $\mathbf{x}$ be a point of X, and let A be a subset of X. Then $\mathbf{x}$ is near A if and only if condition (2.22) holds.*

Proof. Suppose that near $(\mathbf{x}, A)$, and consider sets $U_k \subset X_k$ such that $\mathbf{x} \in -U_1 \times -U_2$. By Lemma 2.5.3,

$$\mathbf{x} \in - ((U_1 \times X_2) \cup (X_1 \times U_2)),$$

so there exists $\mathbf{y}$ in

$$A - ((U_1 \times X_2) \cup (X_1 \times U_2)).$$

Again applying Lemma 2.5.3, we see that this last set equals $(-U_1 \times -U_2) \cap A$.

Conversely, suppose that (2.22) holds, and consider any subset B of X such that $\mathbf{x} \bowtie B$. There exist sets $U_k \subset X_k$ such that $\mathbf{x} \in -U_1 \times -U_2 \subset \sim B$. Then $-U_1 \times -U_2 \subset -B$, so

$$A - B \supset (-U_1 \times -U_2) \cap A,$$

which, by our hypothesis (2.22), is inhabited. Hence

$$\forall_{B \subset X} (\mathbf{x} \bowtie B \Rightarrow \exists_{\mathbf{y}} (\mathbf{y} \in A - B))$$

—that is, near$(\mathbf{x}, A)$. ∎

The following is another way of looking constructively at the classical equivalence of near $(\mathbf{x}, A)$ and $\neg (\mathbf{x} \bowtie A)$.

Proposition 2.5.9 *Let* $X \equiv X_1 \times X_2$ *be the product of two pre-apartness spaces, let* $\mathbf{x} \in X$, *and let* $A \subset X$. *Suppose that the following condition holds:*

(*)   *There exist* $V_1 \subset X_1$ *and* $V_2 \subset X_2$ *such that* $\mathbf{x} \in -V_1 \times -V_2$ *and* $A \subset (V_1 \times X_2) \cup (X_1 \times V_2)$.

Then $\mathbf{x} \bowtie A$. *Conversely, if the spaces* X_1, X_2 *are locally decomposable and* $\mathbf{x} \bowtie A$, *then condition* (*) *holds.*

Proof. First assume (*) and construct V_1, V_2 with the stated properties. We have

$$-V_1 \times -V_2 \subset {\sim}V_1 \times {\sim}V_2$$
$$= ({\sim}V_1 \times X_2) \cap (X_1 \times {\sim}V_2)$$
$$= {\sim}(V_1 \times X_2) \cap {\sim}(X_1 \times V_2)$$
$$= {\sim}((V_1 \times X_2) \cup (X_1 \times V_2)).$$

Hence

$$\mathbf{x} \in -V_1 \times -V_2 \subset {\sim}A$$

and therefore $\mathbf{x} \bowtie A$.

Now assume, conversely, that the spaces X_1, X_2 are locally decomposable and that $\mathbf{x} \bowtie A$. Choose $U_k \subset X_k$ such that

$$\mathbf{x} \in -U_1 \times -U_2 \subset {\sim}A.$$

For each k, use the local decomposability of X_k to find $V_k \subset X_k$ such that $x_k \in -V_k$ and $X_k = -U_k \cup V_k$. Then

$$A \subset (V_1 \times X_2) \cup (X_1 \times V_2).$$

For if $(a_1, a_2) \in A$, then either $a_1 \in -U_1$ and $a_2 \in -U_2$, which is impossible, or else, as must be the case, $a_1 \in V_1$ or $a_2 \in V_2$. Thus (*) holds. ∎

We now look at the continuity of mappings into and out of product spaces. It is natural to begin with the continuity of the projection mappings on a product pre-apartness space. For this we need

Proposition 2.5.10 *Let* $X \equiv X_1 \times X_2$ *be a product of pre-apartness spaces, and let* $A_k \subset X_k$ $(k = 1, 2)$. *Then* $A_1 \times A_2$ *is nearly open in* X *if and only if, for each* k, A_k *is nearly open in* X_k.

Proof. Suppose first that $A_1 \times A_2$ is nearly open in X, and let $\mathbf{x} \in A_1 \times A_2$. Then there exists $S \subset X$ such that $\mathbf{x} \in -S \subset A_1 \times A_2$. Construct $U_k \subset X_k$ such that $\mathbf{x} \in -U_1 \times -U_2 \subset {\sim}S$; then

$$\mathbf{x} \in -U_1 \times -U_2 \subset -S \subset A_1 \times A_2,$$

so $x_1 \in -U_1 \subset A_1$ and $x_2 \in -U_2 \subset A_2$. It readily follows that A_k is nearly open in X_k.

The converse is a straightforward application of Lemma 2.5.3. ∎

Corollary 2.5.11 *Let* $X \equiv X_1 \times X_2$ *be a product of pre-apartness spaces. Then the apartness topology on* X *is the product of the apartness topologies on* X_1 *and* X_2.

Proposition 2.5.12 *Let* $X \equiv X_1 \times X_2$ *be a product of pre-apartness spaces. Then for each* k, *the projection mapping* $\mathrm{pr}_k : X \to X_k$ *is topologically continuous. If the apartness topology on* X *has the reverse Kolmogorov property, then* pr_k *is continuous.*

Proof. If $U \subset X_1$, then

$$\mathrm{pr}_1^{-1}(-U) = -U \times X_2 = -U \times -\varnothing,$$

which, by Proposition 2.5.10, is nearly open in X. Hence pr_1, and similarly pr_2, is topologically continuous. For the second conclusion, we need only apply Proposition 2.3.6. ∎

Proposition 2.5.13 *Let* $X \equiv X_1 \times X_2$ *be the product of two pre-apartness spaces, let* $s \equiv (\mathbf{x}_n)_{n \in \mathfrak{D}}$ *be a net in* X, *and let* $\xi \in X$. *Then* s *converges to* ξ *in* X *if and only if, for* $k \in \{1, 2\}$, *the net* $\mathrm{pr}_k \circ s$ *converges to* ξ_k *in* X_k.

Proof. *Only if* follows from Propositions 2.5.12 and 2.4.16. To prove *if*, suppose that, for $k \in \{1, 2\}$, the net $\mathrm{pr}_k \circ s$ converges in X_k to ξ_k. Let $\xi \in -U$ in X, and construct $U_k \subset X_k$ such that $\xi \in -U_1 \times -U_2 \subset -U$. Then $\xi_k \in -U_k$, so there exists $n_k \in \mathfrak{D}$ such that $\mathrm{pr}_k(\mathbf{x}_n) \in -U_k$ for all $n \succcurlyeq n_k$. Compute $N \in \mathfrak{D}$ such that $N \succcurlyeq n_1$ and $N \succcurlyeq n_2$. Then for all $n \succcurlyeq N$ we have $\mathbf{x}_n \in -U_1 \times -U_2 \subset -U$. Hence s converges in X to ξ. ∎

Proposition 2.5.14 *Let* $X \equiv X_1 \times X_2$ *be a product of pre-apartness spaces, and* f *a continuous mapping of* X *into a pre-apartness space* Y. *Then for each* $x_2 \in X_2$ *the mapping* $x \rightsquigarrow f(x, x_2)$ *is continuous on* X_1, *and for each* $x_1 \in X_1$ *the mapping* $x \rightsquigarrow f(x_1, x)$ *is continuous on* X_2.

Proof. Fixing x_2 in X_2, define $g(x) \equiv f(x, x_2)$. Let $x \in X_1$ and $A \subset X_1$ satisfy $g(x) \bowtie g(A)$—that is,

$$f(x, x_2) \bowtie f(A \times \{x_2\}).$$

Since f is continuous, we have $(x, x_2) \bowtie A \times \{x_2\}$, so there exist $U_k \subset X_k$ such that

$$(x, x_2) \in -U_1 \times -U_2 \subset\sim (A \times \{x_2\}).$$

Hence $x \in -U_1 \subset \sim A$, and therefore $x \bowtie A$, by **A4** in X_1. Thus g is continuous. A similar argument shows that $x \rightsquigarrow f(x_1, x)$ is continuous on X_2 for each $x_1 \in X_1$. ∎

The foregoing results enable us now to show that the product of two locally decomposable apartness structures has a desirable categorical property.

Proposition 2.5.15 *Let $X \equiv X_1 \times X_2$ be the product of two locally decomposable apartness spaces, and f a mapping of a pre-apartness space Y into X. Then f is continuous if and only if $\mathrm{pr}_k \circ f$ is continuous for each k.*

Proof. Assume that $\mathrm{pr}_k \circ f$ is continuous for each k. Let $y \in Y$ and $T \subset Y$ satisfy $f(y) \bowtie f(T)$. By Propositions 2.5.6 and 2.5.9, there exist $V_1 \subset X_1$ and $V_2 \subset X_2$ such that $f(y) \in -V_1 \times -V_2$ and

$$f(T) \subset (V_1 \times X_2) \cup (X_1 \times V_2).$$

Setting

$$T_1 \equiv f^{-1}\left(f(T) \cap (V_1 \times X_2)\right),$$
$$T_2 \equiv f^{-1}\left(f(T) \cap (X_1 \times V_2)\right),$$

we have $T \subset T_1 \cup T_2$. Moreover, for each k, $\mathrm{pr}_k \circ f(T_k) \subset V_k$, so

$$\mathrm{pr}_k \circ f(y) \in -V_k \subset -\mathrm{pr}_k \circ f(T_k).$$

Our continuity hypothesis now ensures that $y \in -T_1 \cap -T_2 \subset -T$.

The converse is simple, in view of Proposition 2.5.12. ∎

Since **A5**, and hence the reverse Kolmogorov property, is a consequence of local decomposability, why did we not simply require that every apartness space be locally decomposable? We could have done this, but we chose not to in order to base our development on the weakest possible principles, with the application of stronger ones, such as **A5** or local decomposability, clearly signalled when it is needed. Nevertheless, local decomposability will play a significant part in what follows in Chapter 3—namely, the theory of apartness between sets.

The insomniac reader will have noticed that, although we have discussed the flow of several properties between a product space and its factors, we have not said anything in this connection about the reverse Kolmogorov property.

Proposition 2.5.16 *If the product $X \equiv X_1 \times X_2$ of two pre-apartness spaces has the reverse Kolmogorov property, then so do the spaces X_k. Conversely, if one of the spaces X_k has the property **A5** and both have the reverse Kolmogorov property, then X has the reverse Kolmogorov property.*

Proof. Suppose that X has the reverse Kolmogorov property. Consider $x, y \in X_1$ and $U \subset X_1$ such that $x \in -U$ and $y \notin -U$. Fix $b \in X_2$. Then, by Lemma 2.5.2,

$$(x, b) \in -U \times X_2 = -(U \times X_2).$$

On the other hand, if $(y, b) \in -(U \times X_2)$, then $y \in -U$, a contradiction; so $(y, b) \notin -(U \times X_2)$ and therefore, by the reverse Kolmogorov property in X, $(x, b) \neq (y, b)$. It now follows from the definition of the inequality on the product space that $x \neq y$ in X_1. Thus X_1, and similarly X_2, has the reverse Kolmogorov property.

Now suppose that both X_1 and X_2 have the reverse Kolmogorov property and that X_1 is an apartness space. Let $\mathbf{x} \in -S$ and $\mathbf{y} \notin -S$ in X. Pick $U_k \subset X_k$ such that $\mathbf{x} \in -U_1 \times -U_2 \subset \sim S$. Since $x_1 \in -U_1$, by **A5** in X_1 we have either $y_1 \neq x_1$ or $y_1 \in -U_1$. In the latter case, since $\mathbf{y} \notin -U_1 \times -U_2$, we have $y_2 \notin -U_2$; but $x_2 \in -U_2$, so, by the reverse Kolmogorov property in X_2, $x_2 \neq y_2$. Thus in either case, $\mathbf{x} \neq \mathbf{y}$. A similar argument covers the case where X_2 has the property **A5**. ∎

The remarkable thing here is that, in the second part of Proposition 2.5.16, we cannot drop with impunity the hypothesis that one of the factor spaces X_k has the property **A5**; if we do so for even two-point spaces X_k, then we get entangled with the **weak disjunctive version of Markov's principle (MP$_{\text{or}}$)**:

> If $(a_n)_{n \geqslant 1}$ is a binary sequence for which it is impossible that all terms equal 0, then

$$\neg \forall_n \left(a_{2n} = 0 \right) \vee \neg \forall_n \left(a_{2n-1} = 0 \right). \tag{2.23}$$

This statement is a simple consequence of Markov's principle.

Lemma 2.5.17 *The following are equivalent.*

(i) **MP$_{\text{or}}$**.

(ii) *If x_1, x_2 are real numbers such that*

$$\neg \left(x_1 = 0 \wedge x_2 = 0 \right), \tag{2.24}$$

then

$$\neg \left(x_1 = 0 \right) \vee \neg \left(x_2 = 0 \right).$$

Proof. Assuming **MP$_{\text{or}}$**, let x_1, x_2 be real numbers such that (2.24) holds. Compute an increasing binary sequence $(\lambda_n)_{n \geqslant 1}$ such that

$$\lambda_n = 0 \Rightarrow \max \left\{ |x_1|, |x_2| \right\} < \frac{1}{n},$$

$$\lambda_n = 1 - \lambda_{n-1} \Rightarrow \max\{|x_1|, |x_2|\} > \frac{1}{n+1}.$$

We may assume that $\lambda_1 = 0$. If $\lambda_n = 0$ or $\lambda_{n-1} = 1$, set $a_{2n} = a_{2n-1} \equiv 0$. If $\lambda_n = 1 - \lambda_{n-1}$, pick $k \in \{1,2\}$ such that $|x_k| > 1/(n+1)$; if $k = 1$, set $a_{2n} \equiv 0, a_{2n-1} \equiv 1$; and if $k = 2$, set $a_{2n} \equiv 1, a_{2n-1} \equiv 0$. Consider the resulting sequence $(a_n)_{n \geqslant 1}$. If $a_n = 0$ for all n, then $\lambda_n = 0$ for all n, so $x_1 = 0 = x_2$, which contradicts (2.24). Thus we can apply **MP$_{\text{or}}$**, to obtain (2.23). If the first alternative in (2.23) holds, suppose that $x_2 = 0$. If

$\lambda_k = 1 - \lambda_{k-1}$, then we must have $|x_1| > 1/(k+1)$ and $a_{2n} = 0$ for all n, a contradiction. Hence $\lambda_n = 0$ for all n, a further contradiction, from which we conclude that $\neg\,(x_2 = 0)$. Likewise, if the second alternative in (2.23) holds, then $\neg\,(x_1 = 0)$. Thus (i) implies (ii).

To prove the converse, assume (ii) and consider any binary sequence $(a_n)_{n\geqslant 1}$ with not all terms equal to 0. The real numbers

$$x_1 \equiv \sum_{n=1}^{\infty} 2^{-n} a_{2n}, \quad x_2 \equiv \sum_{n=1}^{\infty} 2^{-n} a_{2n-1} \qquad (2.25)$$

satisfy (2.24). Hence either $\neg(x_1 = 0)$ or $\neg(x_2 = 0)$, from which we obtain (2.23). ∎

By the **denial topology** on an inhabited set X we mean the topology τ' for which the logical complements of subsets of X form a base of open sets.

Proposition 2.5.18 *The following statement is equivalent to* $\mathrm{MP_{or}}$:

(†)   *For each pair (x_1, x_2) of real numbers, if $X_k \equiv \{0, x_k\}$ is given the denial inequality and topology, then the product topological space $X_1 \times X_2$ has the reverse Kolmogorov property.*

Proof. Assuming (†), let x_1, x_2 be real numbers such that (2.24) holds, and let $X_k \equiv \{0, x_k\}$ be given the denial inequality and topology. Define

$$U_k \equiv \{x \in X_k : \neg\,(x = x_k)\}.$$

Then $x_k \in \neg U_k$, so (x_1, x_2) belongs to the open subset $\neg U_1 \times \neg U_2$ of the product topological space $X \equiv X_1 \times X_2$. On the other hand, if $(0,0) \in \neg U_1 \times \neg U_2$, then $\neg\neg\,(0 = x_1)$ and $\neg\neg\,(0 = x_2)$, which implies that

$$\neg\neg\,(x_1 = 0 \wedge x_2 = 0),$$

a contradiction. Thus $(0,0) \notin \neg U_1 \times \neg U_2$. Since X has the reverse Kolmogorov property, it follows that $(x_1, x_2) \neq (0,0)$; whence

$$\neg\,(x_1 = 0) \vee \neg\,(x_2 = 0).$$

Thus (†) implies statement (ii) of Lemma 2.5.17 and therefore, by that lemma, $\mathrm{MP_{or}}$.

Now suppose, conversely, that $\mathrm{MP_{or}}$ holds. Let $x_1, x_2 \in \mathbf{R}$, let $X_k \equiv \{0, x_k\}$ be given the denial inequality and topology, and let $X \equiv X_1 \times X_2$ be the corresponding product topological space. Consider $\xi, \eta \in X$ and an open set $U \subset X$ such that $\xi \in U$ and $\eta \notin U$. We want to prove that $\xi \neq \eta$ in X—in other words, that there exists $k \in \{1, 2\}$ such that $\neg\,(\xi_k = \eta_k)$. Construct sets $U_k \subset X_k$ such that $\xi \in \neg U_1 \times \neg U_2 \subset U$; then $\eta \notin \neg U_1 \times \neg U_2$. Hence

$$\neg\,(\xi_1 - \eta_1 = 0 \wedge \xi_2 - \eta_2 = 0);$$

so, by Lemma 2.5.17, there exists $k \in \{1,2\}$ such that $\neg (\xi_k - \eta_k = 0)$ and therefore $\neg (\xi_k = \eta_k)$. Hence $\xi \neq \eta$ in X, and we have proved that $\mathbf{MP_{or}}$ implies (†). ∎

For any inhabited set with the denial inequality, the pre-apartness induced by the denial topology τ_k' is just the **denial pre-apartness**, given by

$$x \bowtie' S \Leftrightarrow x \in \neg S,$$

and the space (X, τ') is topologically consistent.

Corollary 2.5.19 $\mathbf{MP_{or}}$ is equivalent to the statement: For each pair (x_1, x_2) of real numbers, if $X_k \equiv \{0, x_k\}$ is given the denial inequality and pre-apartness, then the product pre-apartness space has the reverse Kolmogorov property.

Proof. This follows from Propositions 2.5.18 and 2.2.11, with reference to the remark immediately preceding this corollary. ∎

2.6 Concluding Remarks on Impredicativity

There is one serious issue that needs to be addressed before we conclude this chapter: namely, the quantification over subsets of X in the definition of the expression near (x, A). Such quantification appears to allow impredicativity, a notion viewed with horror by many of the pioneers of constructivism. Indeed, Beeson ([6], page 19) and others have suspected that the power set axiom—implicit in a second-order theory like ours as it stands—is inherently nonconstructive; to quote Myhill [72],

> Power set seems especially nonconstructive and impredicative ... it does not involve ... putting together or taking apart sets that one has already constructed but rather selecting, out of the totality of all sets, all those that stand in the relation of inclusion to a given set.

Is it, then, the case that our theory really does depend on the full power-set axiom? The answer, we believe, is a firm "no".

To justify this belief, we first note that the quantification in our definition of near (x, A) takes place, not over all subsets of X, but *over subsets of X that are apart from x*:

$$\text{near}\,(x, A) \Leftrightarrow \forall_{S \subseteq X} \,(x \bowtie S \Rightarrow \exists_{y \in A} \,(y \bowtie S)).$$

Therefore, in order to avoid impredicativity, we need only that for each x in X the set of all subsets of X that are apart from x be well defined constructively. This can be accomplished by prescribing, for each inhabited set X with an inequality relation, a suitable family $\mathcal{A}(X)$ of subsets to which, and to no others, the relation $\bowtie$ may be applied. Such a family would need to satisfy at least (and maybe only) the following conditions:

- $\{x\} \in \mathcal{A}(X)$ for each $x \in X$.

- Finite unions of sets in $\mathcal{A}(X)$ belong to $\mathcal{A}(X)$.

- For any map $f : X \to Y$ between sets with inequality relations, if $S \in \mathcal{A}(X)$, then $f(S) \in \mathcal{A}(Y)$.

- For any map $f : X \to Y$ between sets with inequality relations, if $T \in \mathcal{A}(Y)$, then $f^{-1}(T) \in \mathcal{A}(X)$.

- Any interval belongs to $\mathcal{A}(\mathbf{R})$.

We would then impose upon the relation $\bowtie$ the requirement that $x \bowtie A$ only if $A \in \mathcal{A}(X)$. So, for example, axiom **A4** would then be written more carefully as

$$\forall_{A \in \mathcal{A}(X)} \forall_{B \in \mathcal{A}(X)} \left(-A \subset \sim B \Rightarrow -A \subset -B \right).$$

Another way of dealing with the spectre of impredicativity is based on the observation that some authors—for example Myhill [72]—while rejecting the power-set axiom, are prepared to accept that the set $\mathbf{R}^X$ of strongly extensional mappings from a set X (with an inequality) to $\mathbf{R}$ is constructively well defined. With this in mind, we could adopt the following procedure. First, introduce the axioms **A1**–**A5** for the apartness on X and define the notion of continuity of a mapping between apartness spaces. Next, having defined the canonical apartness on $\mathbf{R}$, introduce an extra axiom which entails that all apartness spaces are completely regular. Then near (x, A) translates as

$$\forall_{f \in [0,1]^X} ((f \text{ is continuous} \land f(x) = 0 \land f(A) \subset \{1\})$$
$$\Rightarrow \exists_{y \in A} (f(y) < 1)),$$

where the universal quantification is taken over a genuine set of mappings.

The disadvantage of the approach suggested in the preceding paragraph is that it confines us to completely regular spaces. On the other hand, complete regularity applies widely in the classical theory of proximity spaces [75], and, as is easily shown, implies local decomposability in our theory.

There is yet another way in which we might avoid impredicativity: namely, by working with an informal notion of *well-constructed subset* of X, intended to capture the idea of a subset of X built up from some basic collection of subsets by purely predicative means. Our definition of near (x, A) would then be re-cast as

$$\forall_S ((S \text{ is a well-constructed subset of } X \land x \bowtie S) \Rightarrow \exists_y (y \in A - S)).$$

This idea is not as imprecise as may at first appear: there is a constructive formalisation of Morse set theory [14] with a universal class (*not* a set) U, in which objects appear to be constructed predicatively if and only if they can be proved to belong to U. In the framework of that set theory, we could solve our impredicativity problem by introducing the following extra axiom for an apartness relation on X:

A0 $x \bowtie S \Rightarrow (x \in X \wedge S \subset X \wedge S \in U)$.

As a last word on impredicativity, we add that, in practice, nearness does not have much of a role to play in our theory; it is apartness that carries the significant computational information about points and sets. Thus little might be lost were we to exclude nearness from our constructive deliberations altogether.

Notes on Chapter 2

The relation between pre-apartness spaces and those that also satisfy **A5** is analogous to that between groups in general and abelian groups.

If Y is an inhabited subset of a pre-apartness space X, then classically the relation $\bowtie_Y$ always satisfies **A4**. For if $A,B \subset Y$ and $Y - A \subset Y \sim B$, then for any $x \in X - A$ we have either $x \in Y - A$ or $x \notin Y$; in either case this entails $x \in X \sim B$. Thus $X - A \subset X \sim B$, and therefore, by **A4** in X, $X - A \subset X - B$; whence $Y - A \subset Y \sim B$.

In the case where $\sim A$ and $\neg A$ coincide, we can prove classically, by contraposition, that if $\neg (x \bowtie A)$, then near (x, A). For if $\neg$near (x, A), then there exists $S \subset X$ such that $x \in -S$ and $A - S = \varnothing$; whence $x \in -S \subset \neg A = \sim A$, and therefore, by **A4**, $x \bowtie A$. Referring to Proposition 2.1.14, we see that in this case, near (x, A) if and only if $\neg (x \bowtie A)$.

The reverse Kolmogorov property was introduced by Ishihara, who gave it the label $\mathbf{T}_0^{\neq}$. It looks rather like a reversal of the standard Kolmogorov property for a topological space—hence its name. For more on that and other properties of an inequality on pre-apartness and related spaces, see [48]. If every pre-apartness space with the reverse Kolmogorov property has the property **A5**, then we can prove **WLPO**. To see this, equip $X \equiv [0,1]$ with the denial inequality and pre-apartness. Then X has the reverse Kolmogorov property. Let $(a_n)_{n \geq 1}$ be any binary sequence, let

$$y \equiv \sum_{n=1}^{\infty} 2^{-n} a_n \in X,$$

and let $S \equiv (0, 1]$. Then $0 \in -S$; but if we have either $0 \neq y$ or $y \bowtie S$, then

$$\neg \forall_n (a_n = 0) \vee \forall_n (a_n = 0).$$

Among the many merits of local decomposability are these: in its presence, continuity and topological continuity coalesce, a topological apartness space is topologically consistent, and, according to Proposition 2.5.15, product apartness spaces have precisely the categorical property that one would wish for.

We are grateful to Jeremy Clarke for the example showing that if every $\mathbf{T}_1$ topological apartness space is topologically consistent, then the law of excluded middle holds. In the absence of topological consistency, there is an interesting range of topologies that can lie between a given one and its related apartness topology [59].

Proposition 2.2.12 and the comments preceding it give us another proof that every metric space (X, ρ) is topologically consistent.

It is an interesting exercise to show that even in the apartness space $\mathbf{R}$ we cannot prove that, in every case, the union of two nearly closed sets is nearly closed.

We do not know whether Proposition 2.3.7 holds without the hypothesis that Y have the weak nested neighbourhoods property (it certainly does under classical logic).

The classical theory of nets requires a partial order where we use a preorder. A partial order is a preorder R with the additional property of antisymmetry:

$$\forall_{x,y} \left((xRy \wedge yRx) \Rightarrow x = y \right).$$

If we used a partial order in our constructive theory of nets, we would run into difficulties which the classical theory avoids by applications of the axiom of choice.

A Kripke model based on that on pages 137–138 of [24] shows that $\mathbf{MP}_{or}$ is independent of **BISH**.

In the first part of the proof of Proposition 2.5.18 we used the logical theorem

$$\neg\neg P \wedge \neg\neg Q \Rightarrow \neg\neg (P \wedge Q),$$

which the reader is invited to prove informally.

When we began the investigations that eventually led us to apartness spaces, we worked with primitive notions of both nearness and apartness [28]; to a large extent, those ideas have been superseded by our current theory of apartness, based on a relatively succinct set of axioms.

We have not attempted a rigorous justification of the statement that, in constructive Morse set theory, objects are constructed impredicatively if and only if they belong to the universe U, but we have little doubt that such a proof could be produced after a tedious examination of the many cases that would arise.

Hedin [51] has discussed means of developing a purely predicative theory of apartness within Martin-Löf type theory.

3

APARTNESS BETWEEN SETS

... the sense of this apartness, being, by all the current standards, a mental excess, became in its turn a cause of wider separation.

Aldous Huxley

Synopsis

In this—perhaps the most important, and certainly the most technically complex—chapter of the book we extend the notion of point-set (pre-)apartness axiomatically to one of (pre-)apartness between subsets of an inhabited set X. We then study quasi-uniform spaces as an important type of set-set apartness space. In contrast to the counterpart classical theory of proximity spaces, it turns out that the constructive theory of apartness spaces is larger than that of quasi-uniform spaces. In Section 3.3 we explore the connection between strong continuity—the natural set-set extension of the notion of continuity—and uniform continuity. In the next section we introduce and compare various structures that form natural settings for various types of convergence in Y^X. Section 3.5 introduces totally Cauchy nets in pre-apartness spaces, and develops a number of highly technical results relating totally Cauchy and uniformly Cauchy nets, and corresponding notions of completeness, in a uniform space. The section ends with a uniform-space analogue of Bishop's lemma for locatedness, which leads us neatly into a section covering almost locatedness, a notion that appears to be almost as powerful as the standard one of locatedness for metric spaces. We then construct the product of two apartness spaces, showing how various (but not all) important properties pass between a product space and its factors. This prepares us for a section that deals with proximal connectedness. The penultimate section of the chapter shows how, with a given apartness, we can produce an associated structure that is almost a uniform one (classically, it is a totally bounded uniform structure) and that has a number of interesting properties; in particular, using this structure, we obtain a positive concept of nearness/proximity of sets. The final section deals with Diener's approach to compactness in apartness spaces, using his notions of *neat locatedness* and *neat compactness*.

3.1   Set-Set Pre-apartness Relations

When we need to clarify the distinction between a pre-apartness between sets and a pre-apartness between points and sets, we shall refer to the former as a *set-set pre-apartness* and to the latter as a *point-set pre-apartness*. In this chapter we introduce set-set pre-apartness and thereby obtain a much richer theory than that between points and sets; indeed, the set-set theory subsumes the latter, since every set-set pre-apartness has a natural associated point-set pre-apartness to which we can apply the results in Chapter 2.

As before, X will be an inhabited set with an inequality, but this time it will be equipped with a **pre-apartness** relation $\bowtie$ between pairs of subsets of X. Defining

$$-S \equiv \{x \in X : \{x\} \bowtie S\} \tag{3.1}$$

(a definition that, as will be clear shortly, ties in perfectly with its counterpart in Chapter 2), we require that $\bowtie$ satisfy the following four axioms:

B1 $X \bowtie \varnothing$

B2 $-A \subset \sim A$

B3 $((A_1 \cup A_2) \bowtie (B_1 \cup B_2)) \Leftrightarrow \forall_{i,j\in\{1,2\}} A_i \bowtie B_j$

B4 $-A \subset \sim B \Rightarrow -A \subset -B.$

We then call the pair $(X,\bowtie)$, or when no confusion is likely, the set X itself, a **pre-apartness space**; and if $A \bowtie B$, we say that A is **apart from** B. Note that, at this stage, we do not require that the relation $\bowtie$ be **symmetric**: that is,

$$\forall_{A,B\subset X} (A \bowtie B \Leftrightarrow B \bowtie A). \tag{3.2}$$

If the relation is symmetric and $A \bowtie B$, then we say that the sets A,B are **apart** (from each other).

Defining

$$x \bowtie A \Leftrightarrow \{x\} \bowtie A, \tag{3.3}$$

we obtain a point-set pre-apartness associated with the given set-set one: it is trivial to show that axioms **A1–A3** are satisfied by the point-set relation $\bowtie$; the point-set axiom **A4** is precisely our set-set axiom **B4**. From now on we think of a set-set pre-apartness space X also as a point-set apartness space with the point-set apartness defined at (3.3).

The canonical example of a set-set pre-apartness space is a quasi-metric space (X, ρ), where the apartness between subsets A, B is defined by

$$A \bowtie B \Leftrightarrow \exists_{r>0} \forall_{x \in A} \forall_{y \in B} \left(\rho(x, y) \geqslant r \right).$$

We shall generalise this to uniform spaces in Section 3.2.

Lemma 3.1.1 *If S, T are subsets of a pre-apartness space $(X, \bowtie)$ and $S \bowtie T$, then $A \bowtie B$ for all $A \subset S$ and $B \subset T$.*

Proof. Let $A \subset S$ and $B \subset T$. Then $S = A \cup S$ and $T = B \cup T$, so $A \cup S \bowtie B \cup T$. The result now follows from **B3**. ∎

Proposition 3.1.2 *In any pre-apartness space X, $\varnothing \bowtie \varnothing$.*

Proof. Apply **B1** and Lemma 3.1.1. ∎

Lemma 3.1.3 *If A, B are subsets of a pre-apartness space $(X, \bowtie)$ and $A \bowtie B$, then $A \subset -B$ and $B \subset {\sim} A$.*

Proof. For each $x \in A$ we have $\{x\} \bowtie B$—that is, $x \in -B$—by Lemma 3.1.1. Hence $A \subset -B$. It follows from axiom **B2** that $A \subset {\sim}B$; whence $B \subset {\sim}A$. ∎

If X is a set-set pre-apartness space, then (see page 20) $\varnothing = -X$. Moreover, if $A \bowtie X$, then, by Lemma 3.1.3, $A \subset -X \subset \neg X$; whence $A \subset \varnothing$ and therefore $A = \varnothing$. Thus the only subset of X that could be apart from X is the empty subset. Although without the symmetry of $\bowtie$ we cannot prove that $\varnothing \bowtie X$, we can prove the following:

Proposition 3.1.4 *Let T be a subset of a pre-apartness space $(X, \bowtie)$ such that $-T$ is inhabited. Then $\varnothing \bowtie T$.*

Proof. Pick $s \in -T$. Then $\{s\} \bowtie T$ and $\varnothing \subset \{s\}$, so the conclusion follows from Lemma 3.1.1. ∎

By a (set-set) **apartness** on X we mean a relation $\bowtie$ between subsets of X that satisfies axioms **B1–B3** and

B5 $x \in -A \Rightarrow \exists_{S \subset X}(x \in -S \wedge X = -A \cup S).$

It then also satisfies **B4** and so is a pre-apartness. To see this, let $x \in -A \subset {\sim}B$. Then by **B5**, there exists S such that $x \in -S$ and $X = -A \cup S$. Since $B \cap -A = \varnothing$, we have $B \subset S$. It follows from Lemma 3.1.1 that $x \in -B$. Hence $-A \subset -B$.

Regarded as a point-set pre-apartness space, X is then locally decomposable (this is precisely what **B5** says) and hence satisfies **A5**. Thus it makes sense to call $(X, \bowtie)$, or simply X itself, a (set-set) **apartness space**. Every quasi-metric pre-apartness space is an apartness space.

Lemma 3.1.5 *Let X be a symmetric set-set apartness space, $x \in X$, and $A \subset X$. If $x \in -A$, then $X = -\{x\} \cup -A$.*

Proof. By **B5**, there exists $S \subset X$ such that $x \in -S$ and $X = -A \cup S$. If $y \in S$, then $x \bowtie \{y\}$, by Lemma 3.1.1, and therefore, by symmetry, $y \bowtie \{x\}$. Hence $S \subset -\{x\}$ and therefore $X = -\{x\} \cup -A$. ∎

We pause in our main development of set-set apartness to introduce two strong (hence the "s" subscripts) forms of axiom **B4/A4** that hold in many important pre-apartness spaces:

A4$_s$ $(A \bowtie B \wedge -B \subset {\sim}C) \Rightarrow A \bowtie C$,

A4$_{ss}$ $(A \bowtie B \wedge -B \subset \neg C) \Rightarrow A \bowtie C$.

Note that **A4$_{ss}$** implies **A4$_s$**, which in turn implies **A4**.

Proposition 3.1.6 *If X is a set-set pre-apartness space satisfying **A4$_s$**, then for all subsets A and B of X,*

$$A \bowtie B \Leftrightarrow A \bowtie {\sim}{\sim}B. \qquad (3.4)$$

Proof. If $A \bowtie B$, then, since $-B \subset {\sim}B = {\sim}{\sim}{\sim}B$, we see from **A4$_s$** that $A \bowtie {\sim}{\sim}B$. On the other hand, the right-to-left implication in (3.4) holds even without **A4$_s$**, since $B \subset {\sim}{\sim}B$ and Lemma 3.1.1 can be applied. ∎

A similar proof, which we omit, establishes

Proposition 3.1.7 *If X is a set-set pre-apartness space satisfying **A4$_{ss}$**, then for all subsets A and B of X,*

$$A \bowtie B \Leftrightarrow A \bowtie \neg\neg B.$$

For the next proposition, the reader should bear in mind that, although $\neg S \cup \neg T \subset \neg(S \cap T)$, the reverse inclusion does not hold constructively for arbitrary sets S,T.

Proposition 3.1.8 *Let X be a pre-apartness space that satisfies **A5** and **A4$_s$**. Then*

$$\forall_{A,S,T \subset X} \left(A \bowtie (\neg S \cup \neg T) \Leftrightarrow A \bowtie \neg(S \cap T) \right). \qquad (3.5)$$

Proof. Let $A \bowtie (\neg S \cup \neg T)$, let $x \in -(\neg S \cup \neg T)$, and let $y \in \neg(S \cap T)$. By **B3**, $x \bowtie \neg S$ and $x \bowtie \neg T$; so, by two applications of **A5**, either $x \neq y$, or else

$$y \in -\neg S \subset \neg\neg S \text{ and } y \in -\neg T \subset \neg\neg T. \qquad (3.6)$$

We show that (3.6) is ruled out. Suppose, then, that (3.6) holds. If $y \in S$, then, as $y \in \neg(S \cap T)$, we must have $y \in \neg T$, a contradiction; whence $y \in \neg S$, again a contradiction. Thus (3.6) is ruled out, and therefore $x \neq y$.

Since $x \in -(\neg S \cup \neg T)$ and $y \in \neg(S \cap T)$ are arbitrary, we conclude that $-(\neg S \cup \neg T) \subset \sim \neg(S \cap T)$. It follows from **A4**$_s$ that $A \bowtie \neg(S \cap T)$. This proves the left-to-right implication in (3.5). In view of the remark preceding this proposition, the right-to-left implication follows from Lemma 3.1.1. ∎

The strongest of all the separation properties normally considered for an apartness space X is the **Efremovič property**:

EF $A \bowtie B \Rightarrow \exists_{E \subset X} (A \bowtie \neg E \wedge E \bowtie B)$.

This, too, is a property that holds in a large variety of apartness spaces. Note that the symbol $\Rightarrow$ in **EF** can be replaced by $\Leftrightarrow$: for if $A \bowtie \neg E$ and $E \bowtie B$, then $E \subset -B \subset \sim B$, by Lemma 3.1.3 and **B2**, so

$$B \subset \sim \sim B \subset \sim E \subset \neg E$$

and therefore, by Lemma 3.1.1, $A \bowtie B$.

Here is an example of a symmetric pre-apartness space in which the Efremovič property holds if and only if we can derive Markov's principle. Define a symmetric pre-apartness $\bowtie$ on $\mathbf{R}$ (taken with its usual inequality relation and the corresponding notion of complement $\sim$) by

$$A \bowtie B \Leftrightarrow A \subset \sim B.$$

Then $0 \bowtie \sim \{0\}$. Suppose there exists $E \subset \mathbf{R}$ such that $0 \bowtie \neg E$ and $E \bowtie \sim \{0\}$. Then $\neg(x \neq 0)$ for each $x \in E$, so, since the inequality on $\mathbf{R}$ is tight, $E \subset \{0\}$ and therefore $\neg \{0\} \subset \neg E$. Hence $0 \bowtie \neg \{0\}$ and therefore, by symmetry, $\neg \{0\} \subset \sim \{0\}$. Thus

$$\forall_{x \in \mathbf{R}} (\neg(x = 0) \Rightarrow x \neq 0),$$

which is equivalent to Markov's principle. Conversely, suppose that Markov's principle holds; then $\sim S = \neg S$ for each $S \subset \mathbf{R}$. If subsets A and B of $\mathbf{R}$ are apart relative to the pre-apartness in question, then

$$A \subset \sim B = \sim \sim \sim B = \sim \neg \neg B$$

and therefore $A \bowtie \neg \neg B$. On the other hand, $\neg B = \sim B$, so $\neg B \bowtie B$. Thus **EF** holds with $E = \neg B$.

Proposition 3.1.9 *A pre-apartness space X with the Efremovič property has the reverse Kolmogorov property.*

Proof. Let $U \subset X$, and let x,y be points of X such that $x \in -U$ and $y \notin -U$. By **EF**, there exists $E \subset X$ such that $x \bowtie \neg E$ and $E \bowtie U$. If $y \in E$, then $y \in -U$, a contradiction. Hence $\{y\} \subset \neg E$; so $x \bowtie \{y\}$, by Lemma 3.1.1, and therefore (by **B2**) $x \neq y$. ∎

Proposition 3.1.10 *For a pre-apartness space X, the Efremovič property implies* **A4**$_{ss}$.

Proof. Let $A \bowtie B$ and $-B \subset \neg C$. By **EF**, there exists $E \subset X$ such that $A \bowtie \neg E$ and $E \bowtie B$. So $E \subset -B \subset \neg C$ and therefore $C \subset \neg\neg C \subset \neg E$. It follows from Lemma 3.1.1 that $A \bowtie C$. ∎

Proposition 3.1.11 *A symmetric* **T**$_1$ *pre-apartness space with the Efremovič property is Hausdorff.*

Proof. Let X be a symmetric **T**$_1$ pre-apartness space, and let $x,y \in X$ and $x \neq y$. Since X is **T**$_1$, $x \bowtie \{y\}$. By **EF** and symmetry, there exists $V \subset X$ such that $x \bowtie \neg V$ and $y \bowtie V$. Taking $U \equiv \neg V$, we now have $x \in -U$, $y \in -V$, and (by **B2**) $-U \subset \sim\neg V \subset \sim -V$. ∎

Proposition 3.1.12 *For a symmetric pre-apartness space X, the Efremovič property implies the weak nested neighbourhoods property.*

Proof. Let $x \in -A$ in the pre-apartness space X. Then $A \bowtie \{x\}$, by symmetry, so by **EF**, there exists $E \subset X$ such that $A \bowtie \neg E$ and $E \bowtie \{x\}$. Then, using symmetry and Lemma 3.1.3, we have $\neg E \subset -A$ and $x \in -E$. ∎

A quasi-metric apartness space (X, ρ) satisfies **EF**. For if $S \bowtie T$ in X, then there exists $\varepsilon > 0$ such that $\rho(s,t) > \varepsilon$ for all $s \in S$ and $t \in T$. Letting

$$E \equiv \left\{ x \in X : \exists_{s \in S} \left(\rho(s, x) < \frac{\varepsilon}{2} \right) \right\},$$

for all $s \in S$ and all $x \in \neg E$ we have $\rho(s,x) \geqslant \varepsilon/2$; whence $S \bowtie \neg E$. On the other hand, if $t \in T$ and $x \in E$, then, choosing $s \in S$ such that $\rho(s,x) < \varepsilon/2$, we see that $\rho(x,t) \geqslant \rho(s,t) - \rho(s,x) > \varepsilon/2$. Hence $E \bowtie T$.

If X is a pre-apartness space, and Y is an inhabited subset of X with the induced inequality (see page 22), we define the relation $\bowtie_Y$ between subsets A,B of Y by

$$A \bowtie_Y B \Leftrightarrow A \bowtie B,$$

where, on the right side, $\bowtie$ is the original pre-apartness on X. We say that $\bowtie_Y$ is **induced** on Y by its counterpart on X. It is easy to show that $\bowtie_Y$ satisfies **B1–B3**. If also,

$$(Y - A \subset Y \sim B) \Rightarrow (Y - A \subset Y - B), \tag{3.7}$$

then $\bowtie_Y$ is a pre-apartness on Y, and Y, taken with the induced inequality and pre-apartness, is called a **pre-apartness subspace** of X. Moreover, if $(Y, \bowtie_Y)$ is locally decomposable, then we call Y an **apartness subspace** of X.

Referring to Proposition 2.1.4 and other work on page 23, we observe the following:

- If the point-set pre-apartness on X has the reverse Kolmogorov property, then (3.7) holds, and Y is a set-set pre-apartness subspace of X with the reverse Kolmogorov property.

- If the pre-apartness on X has property **A5** and hence the reverse Kolmogorov property, then so does $\bowtie_Y$.

Proposition 3.1.13 *Every inhabited subset of a set-set apartness space is an apartness subspace.*

Proof. Let Y be an inhabited subset of an apartness space X. Note that since X is locally decomposable and hence satisfies **A5**, it, and therefore Y, has the reverse Kolmogorov property. It follows from the remarks preceding this proposition that we need only prove that $(Y, \bowtie_Y)$ is locally decomposable. To that end, consider $y \in Y$ and $A \subset Y$ such that $y \bowtie_Y A$. We have $y \bowtie A$ in X, so there exists $S \subset X$ such that $y \bowtie S$ in X, and $X = (X - A) \cup S$. Clearly, $Y = (Y - A) \cup (Y \cap S)$. On the other hand, since $Y \cap S \subset S$, we have $y \bowtie Y \cap S$ in X and therefore $y \bowtie_Y Y \cap S$. ∎

In particular, every inhabited subset Y of a quasi-metric space (X, ρ) is an apartness subspace of X; in that case, the apartness induced on Y by the apartness on X is precisely the quasi-metric apartness on Y considered as a quasi-metric subspace of X.

Proposition 3.1.14 *Let X be a pre-apartness space with the Efremovič property, and let Y be a pre-apartness subspace of X. Then Y has the Efremovič property,*

Proof. Let $S \bowtie_Y T$. Then $S \bowtie T$ in X, so there exists $E \subset X$ such that $S \bowtie \neg E$ and $E \bowtie T$. We now see from Lemma 3.1.1 and the definition of $\bowtie_Y$ that $S \bowtie_Y (\neg E \cap Y)$ and $(E \cap Y) \bowtie_Y T$. Since $\neg E \cap Y = Y \cap \neg (E \cap Y)$, the result follows. ∎

Every topological space (X, τ) has a set-set pre-apartness $\bowtie_\tau$ defined by

$$A \bowtie_\tau B \Leftrightarrow A \subset (\sim B)^\circ. \tag{3.8}$$

Then

$$A \bowtie_\tau B \Rightarrow A \subset (\neg B)^\circ.$$

If the topology τ has the topological reverse Kolmogorov property, then the implication here can be reversed, by Proposition 2.2.2.

Proposition 3.1.15 *Let (X, τ) be a topological space. Then $\bowtie_\tau$, defined at (3.8), is a set-set pre-apartness on X whose associated point-set pre-apartness is the topological one corresponding to τ, and whose associated apartness topology coincides with τ. Moreover, if (X, τ) has the topological reverse Kolmogorov property, then $\bowtie_\tau$ has the Efremovič property.*

Proof. It is straightforward to verify that **B1–B4** hold for $\bowtie_\tau$, that the point-set pre-apartness associated $\bowtie_\tau$ is the topological one corresponding to τ, and that the apartness topology associated with $\bowtie_\tau$ coincides with τ. Suppose also that (X,τ) has the topological reverse Kolmogorov property. Consider subsets A,B of X with $A \bowtie_\tau B$. Setting $E \equiv (\neg B)^\circ$, we have $E \bowtie_\tau B$, by the remark preceding this proposition. On the other hand,

$$A \subset (\neg B)^\circ = E = E^\circ \subset \neg\neg E,$$

so $A \subset (\neg\neg E)^\circ$, and therefore, by the remark preceding this proposition, $A \bowtie_\tau \neg E$. ∎

We already know that every set-set pre-apartness induces a point-set one. The next result shows how to concoct a set-set pre-apartness from a given point-set one.

Proposition 3.1.16 *Let $(X,\bowtie)$ be a point-set pre-apartness space. Then*

$$A \bowtie' B \Leftrightarrow A \subset -B \tag{3.9}$$

defines a set-set pre-apartness on X whose associated point-set pre-apartness is the original one. Moreover, if $\bowtie$ has the reverse Kolmogorov property, then it has the Efremovič property.

Proof. Since the apartness complement $-'$ relative to $\bowtie'$ coincides with the apartness complement $-$ relative to $\bowtie$, it is routine to verify that $\bowtie'$ satisfies axioms **B1–B4** and that the point-set apartness relations $\bowtie$ and $\bowtie'$ coincide. Next suppose that $\bowtie$ has the reverse Kolmogorov property, and let $A \bowtie' B$. Setting $E \equiv -B$, we immediately obtain $E \bowtie' B$. On the other hand, by Proposition 2.1.5 and the reverse Kolmogorov property, we have

$$A \subset -B = -(\sim - B) = -(\neg - B),$$

so $A \bowtie' \neg E$. ∎

It might be suggested that, in order to produce a good theory of apartness between sets, we could dispense with axioms **B1–B4**, take a point-set apartness space, and construct the set-set pre-apartness as in the previous proposition. However, this procedure would not suffice to capture the desired notion of apartness in a metric space, where two sets are apart if and only if they are bounded away from each other: the interval $(0,1)$ in $\mathbf{R}$ would be apart from $(-\infty,0] \cup [1,\infty)$.

Classically, the pre-apartness defined at (3.8) satisfies

$$A \bowtie B \Leftrightarrow A \cap \overline{B} = \varnothing.$$

The condition on the right-hand side of this equivalence is constructively weaker than its counterpart in (3.8): if the two conditions were equivalent, then we could derive Markov's principle. To see this, let $a \in \mathbf{R}$ be such that $\neg\,(a = 0)$. Take $A \equiv \{a\}$ and $B \equiv \{0\}$. Then $A \cap \overline{B} = \varnothing$; but if there exists an open set $U \subset \mathbf{R}$ such that $a \in U \subset \sim B$, then $a \neq 0$.

The next lemma will lead us to a second interesting consequence of $\mathbf{A4}_s$.

Lemma 3.1.17 *Let X be a point-set pre-apartness space with the reverse Kolmogorov property, and let $S \subset X$. Then $-S \subset \sim\overline{S}$, where the bar denotes closure with respect to the apartness topology.*

Proof. Let $x \in -S$ and $y \in \overline{S}$. If $y \in -S$, then, by definition of $\overline{S}$, $-S \cap S$ is inhabited, which is absurd. Hence $y \notin -S$ and, by the reverse Kolmogorov property, $x \neq y$. Thus $-S \subset \sim\overline{S}$. ∎

Proposition 3.1.18 *Let X be a pre-apartness space that satisfies $\mathbf{A4}_s$ and has the reverse Kolmogorov property. Then for all subsets A and B of X,*

$$A \bowtie B \Leftrightarrow A \bowtie \overline{B}.$$

Proof. The right-to-left implication follows from Lemma 3.1.1. For the left-to-right one, assume that $A \bowtie B$. By Lemma 3.1.17, $-B \subset \sim\overline{B}$; whence, by $\mathbf{A4}_s$, $A \bowtie \overline{B}$. ∎

Corollary 3.1.19 *Let X be a symmetric set-set pre-apartness space that satisfies $\mathbf{A4}_s$ and has the reverse Kolmogorov property. Then for all subsets A and B of X,*

$$A \bowtie B \Leftrightarrow \overline{A} \bowtie \overline{B}.$$

Proof. Again, Lemma 3.1.1 provides us with the right-to-left implication. For the reverse one, let $A \bowtie B$ and observe that, by Proposition 3.1.18, $A \bowtie \overline{B}$. The symmetry of $\bowtie$ now yields $\overline{B} \bowtie A$, from which, with one more application of Proposition 3.1.18, we obtain $\overline{B} \bowtie \overline{A}$. A final application of symmetry gives $\overline{A} \bowtie \overline{B}$. ∎

A classically important proximity on a topological space has a symmetric apartness given by

$$A \bowtie B \Leftrightarrow \overline{A} \cap \overline{B} = \varnothing. \tag{3.10}$$

In the context of the apartness topology, we now consider a classically equivalent but constructively stronger version of this apartness. In passing, this shows how to manufacture a symmetric set-set pre-apartness from a given point-set pre-apartness (cf. Proposition 3.1.16).

Proposition 3.1.20 *Let $(X, \bowtie_0)$ be a point-set pre-apartness space, and for $S,T \subset X$ define*

$$S \bowtie T \Leftrightarrow \forall_{x \in X}\,(x \bowtie_0 S \vee x \bowtie_0 T). \tag{3.11}$$

Then $\bowtie$ is a symmetric set-set pre-apartness on X which satisfies $\mathbf{A4}_s$ and whose associated point-set pre-apartness is $\bowtie_0$.

Proof. Since, by **A2**, for a point $x \in X$ we cannot have $x \bowtie_0 \{x\}$, we see that $x \bowtie A$ if and only if $x \bowtie_0 A$. It follows that $\bowtie$ satisfies **B2**. Property **B1** for $\bowtie$ readily follows from **A1** for $\bowtie_0$. Next,

$$
\begin{aligned}
A \bowtie (B \cup C) &\Leftrightarrow \forall_{x \in X} (x \bowtie_0 A \vee x \bowtie_0 (B \cup C)) \\
&\Leftrightarrow \forall_{x \in X} (x \bowtie_0 A \vee (x \bowtie_0 B \wedge x \bowtie_0 C)) \quad \text{by } \mathbf{A3} \text{ for } \bowtie_0 \\
&\Leftrightarrow \forall_{x \in X} ((x \bowtie_0 A \vee x \bowtie_0 B) \wedge (x \bowtie_0 A \vee x \bowtie_0 C)) \\
&\Leftrightarrow \forall_{x \in X} ((x \bowtie_0 A \vee x \bowtie_0 B)) \wedge \forall_{x \in X} (x \bowtie_0 A \vee x \bowtie_0 C) \\
&\Leftrightarrow (A \bowtie B) \wedge (A \bowtie C).
\end{aligned}
$$

Since the relation $\bowtie$ is obviously symmetric, we now see that

$$
\begin{aligned}
(A_1 \cup A_2) \bowtie (B_1 \cup B_2) &\Leftrightarrow ((A_1 \cup A_2) \bowtie B_1) \wedge ((A_1 \cup A_2) \bowtie B_2) \\
&\Leftrightarrow (B_1 \bowtie (A_1 \cup A_2)) \wedge (B_2 \bowtie (A_1 \cup A_2)) \\
&\Leftrightarrow (B_1 \bowtie A_1) \wedge (B_1 \bowtie A_2) \\
&\qquad\qquad \wedge (B_2 \bowtie A_1) \wedge (B_2 \bowtie A_2) \\
&\Leftrightarrow \forall_{i,j \in \{1,2\}} (A_i \bowtie B_j).
\end{aligned}
$$

This completes the proof of **B3** for $\bowtie$. Since **B4** for $\bowtie$ is just **A4** for $\bowtie_0$, we have proved that $\bowtie$ is a set-set pre-apartness on X. Finally, to dispose of $\mathbf{A4}_s$, let $A \bowtie B$ and $-B \subset \sim C$ (bearing in mind the coincidence of the point-set pre-apartness relations, and hence of the apartness complements, under consideration). For each $x \in X$, either $x \bowtie_0 A$ or else $x \bowtie_0 B$. In the latter case, $x \in -B \subset \sim C$ and therefore, by **A4**, $x \bowtie_0 C$. Since $x \in X$ is arbitrary, we have $A \bowtie C$. ∎

We make the transition from this section to the next by introducing what will turn out to be a special case of the apartness that is discussed in more generality and detail in the latter section.

A **topological group** consists of a group G with an inequality relation $\neq$ and a topology τ, such that the mapping $\phi : (x, y) \rightsquigarrow x^{-1}y$ is strongly extensional and (in the usual topological sense—see page 16) continuous relative to the product topology on $G \times G$. In that case, the group multiplication and inversion mappings are (topologically) continuous. To see that they are also strongly extensional, denoting the identity of G by e, we have

$$
\begin{aligned}
x^{-1} \neq y^{-1} &\Rightarrow x^{-1}e \neq y^{-1}e \\
&\Rightarrow x \neq y \vee e \neq e \\
&\Rightarrow x \neq y,
\end{aligned}
$$

the second implication following from the strong extensionality of the mapping ϕ. Likewise,

$$
\begin{aligned}
xy \neq ab &\Rightarrow \left(x^{-1}\right)^{-1} y \neq \left(a^{-1}\right)^{-1} b \\
&\Rightarrow x^{-1} \neq a^{-1} \vee y \neq b \\
&\Rightarrow x \neq a \vee y \neq b \quad \text{(by the foregoing)} \\
&\Rightarrow (x, y) \neq (a, b) .
\end{aligned}
$$

Incidentally, it now follows that $x \neq y$ if and only if $x^{-1}y \neq e$.

For subsets U, V of G we write

$$
UV \equiv \{uv : u \in U, v \in V\} .
$$

If $x \in G$, then we write Ux and xU, rather than $U\{x\}$ and $\{x\}U$.

If $U \in \tau$ and $x \in G$, then xU, being the inverse image of U under the continuous mapping $g \rightsquigarrow x^{-1}g$, is in τ. It follows that the neighbourhoods of x in the topological group G are the sets of the form xU with U a neighbourhood of e. We denote the set of neighbourhoods of x in G by $\mathfrak{N}(x)$. Note that if U is a neighbourhood of e in G, then, as the inversion mapping is continuous on G,

$$
U^{-1} \equiv \left\{x \in G : x^{-1} \in U\right\} \in \mathfrak{N}(e).
$$

We define the **left pre-apartness** $\bowtie^{(L)}$ for subsets S, T of a topological group G by

$$
S \bowtie^{(L)} T \Leftrightarrow \exists_{U \in \mathfrak{N}(e)} \left(S^{-1}T \subset {\sim}U\right), \tag{3.12}
$$

which is equivalent to

$$
S \bowtie^{(L)} T \Leftrightarrow \exists_{U \in \mathfrak{N}(e)} \left(SU \subset {\sim}T\right).
$$

We must prove that (3.12) does, indeed, define a pre-apartness on G.

Proposition 3.1.21 *Let (G, τ) be a topological group. Then $\bowtie^{(L)}$ is a symmetric set-set pre-apartness on G, whose corresponding point-set pre-apartness is just that associated with the topology τ.*

Proof. First note that if $U \in \mathfrak{N}(e)$, S and T are subsets of G, and $S^{-1}T \subset {\sim}U$, then for $u \in U$, $s \in S$, and $t \in T$ we have $s^{-1}t \neq u$ and therefore $t^{-1}s \neq u^{-1}$. Hence $T^{-1}S \subset {\sim}U^{-1}$; since $U^{-1} \in \mathfrak{N}(e)$, we have $T \bowtie^{(L)} S$. Thus $\bowtie^{(L)}$ is symmetric.

Since $G \in \mathfrak{N}(e)$ and $G^{-1}\varnothing = \varnothing = {\sim}G$, axiom **B1** holds for $\bowtie^{(L)}$. Let $\{x\} \bowtie^{(L)} A$ in G; then there exists $U \in \mathfrak{N}(e)$ such that $x^{-1}A \subset {\sim}U$. It follows that for each $y \in A$, $x^{-1}y \neq e$ and therefore $x \neq y$. Hence $-A \subset {\sim}A$, and **B2** holds. In view of the symmetry of $\bowtie^{(L)}$, to verify **B3** it is enough to show that

$$
A \bowtie^{(L)} (B \cup C) \Leftrightarrow A \bowtie^{(L)} B \wedge A \bowtie^{(L)} C. \tag{3.13}
$$

Suppose, then, that $A \bowtie^{(L)} B$ and $A \bowtie^{(L)} C$. There exist $U, V \in \mathfrak{N}(e)$ such that $A^{-1}B \subset {\sim}U$ and $A^{-1}C \subset {\sim}V$. Then $U \cap V$ is a neighbourhood of e, and

$$A^{-1}(B \cup C) = A^{-1}B \cup A^{-1}C \subset {\sim}U \cup {\sim}V \subset {\sim}(U \cap V). \qquad (3.14)$$

Hence $A \bowtie^{(L)} (B \cup C)$. The implication from left to right in (3.13) is straightforward to prove.

Now let $\bowtie_\tau$ be the point-set pre-apartness associated with τ. Given $x \in G$ with $\{x\} \bowtie^{(L)} A$, we can find a τ-open neighbourhood U of e such that $x \in xU \subset {\sim}A$. Since $xU \in \tau$, we see that $x \bowtie_\tau A$. Conversely, if $x \bowtie_\tau A$ and V is a τ-open set such that $x \in V \subset {\sim}A$, then there exists $U \in \mathfrak{N}(e)$ such that $xU \subset V \subset {\sim}A$; so $\{x\} \bowtie^{(L)} A$. Thus the point-set relations $\bowtie^{(L)}$ and $\bowtie_\tau$ coincide. It follows immediately that $\bowtie^{(L)}$ satisfies axiom **B4**; this completes the proof that $\bowtie^{(L)}$ is a set-set pre-apartness on G. ∎

Lemma 3.1.22 *Let* $W \in \mathfrak{N}(e)$ *and* $x \in G$. *Then* ${\sim}xW = x({\sim}W)$.

Proof. Let $a \in {\sim}W$ and $b \in W$. Then $a \neq b$, so $xa \neq xb$. Hence $x({\sim}W) \subset {\sim}xW$. For the reverse inclusion, let $y \in {\sim}xW$. Then $x^{-1}y \in {\sim}W$, so $y \in x({\sim}W)$. ∎

A topological group G is called **decomposable** if for each neighbourhood U of the identity e in G there exists $V \in \mathfrak{N}(e)$ such that $G = U \cup {\sim}V$.

Lemma 3.1.23 *Let* G *be a decomposable topological group, and let* $U \in \mathfrak{N}(e)$. *Then there exists* $V \in \mathfrak{N}(e)$ *such that* $V = V^{-1}$, $VV \subset U$, *and* $G = U \cup {\sim}V$.

Proof. Since the product mapping p is topologically continuous on $G \times G$ and maps (e,e) to e, there exist neighbourhoods S,T of e in G such that $S \times T \subset p^{-1}(U)$. Then $W \equiv S \cap T$ is a neighbourhood of e in G such that $WW = p(W \times W) \subset p(S \times T) \subset U$. Using the decomposability of G, construct $E \in \mathfrak{N}(e)$ such that $G = W \cup {\sim}E$. Setting $V \equiv E \cap E^{-1}$, we see that $V \in \mathfrak{N}(e)$, $V = V^{-1}$, and $VV \subset EE \subset WW \subset U$. Moreover, for each $x \in G$, either $x \in {\sim}E \subset {\sim}V$ or else $x \in W \subset WW \subset U$. ∎

Proposition 3.1.24 *The following are equivalent conditions on a topological group* (G, τ):

(i) G *is decomposable.*

(ii) $(G, \bowtie^{(L)})$ *is locally decomposable.*

(iii) G *is topologically locally decomposable.*

Proof. Suppose that G is decomposable. Let $S \subset G$ and $x \bowtie^{(L)} S$. Pick $U \in \mathfrak{N}(e)$ such that $xU \subset {\sim}S$. Using Lemma 3.1.23, construct $V \in \mathfrak{N}(e)$

such that $VV \subset U$ and $G = U \cup \sim V$. Then apply the decomposability of G once more, to produce $W \in \mathfrak{N}(e)$ such that $G = V \cup \sim W$. Let $T \equiv \sim xW$; then $x^{-1}T \subset \sim W$ and therefore $x \bowtie^{(L)} T$. For each $y \in G$, either $x^{-1}y \in \sim W$ or $x^{-1}y \in V$. In the first case, Lemma 3.1.22 shows us that $y \in T$. In the second, $y \in xV$ and so

$$yW \subset xVW \subset xVV \subset xU \subset \sim S;$$

whence $y \bowtie^{(L)} S$. Thus $x \in -T$ and $G = -S \cup T$. It follows that $(G, \bowtie^{(L)})$ is locally decomposable. Hence (i) implies (ii).

Now assume (ii). Since, by Proposition 3.1.21, the point-set relations $\bowtie^{(L)}$ and $\bowtie_r$ coincide, we see from Proposition 2.2.14 that G is topologically locally decomposable. Thus (ii) implies (iii). Finally, it is almost trivial that (iii) implies (i). ∎

An immediate consequence of the previous proposition is that, relative to $\bowtie^{(L)}$, a decomposable topological group is a set-set apartness space.

An example of a decomposable topological group, in which the left pre-apartness is therefore an apartness, is the group $\mathrm{GL}(n, \mathbf{R})$ of invertible n-by-n real matrices. In this case, the topology is derived from the metric ρ defined, for matrices $A, B \in \mathrm{GL}(n, \mathbf{R})$, by

$$\rho(A, B) \equiv \sup_{1 \leqslant i, j \leqslant n} |a_{ij} - b_{ij}|,$$

where, for example, a_{ij} is the (i, j)th entry of the matrix A.

3.2 Quasi-uniform Spaces

We are about to introduce a fundamental example of a pre-apartness space, for which we need a number of preliminary definitions.

Let S be a set, and $\mathcal{B}$ be an inhabited set of inhabited subsets of S. We call $\mathcal{B}$ a **filter base** on S if the intersection of two sets in $\mathcal{B}$ contains a set in $\mathcal{B}$. If $\mathcal{B}$ has the following stronger properties, then it is called a **filter** on S:

F1 The intersection of two sets in $\mathcal{B}$ belongs to $\mathcal{B}$.

F2 All supersets of sets in $\mathcal{B}$ belong to $\mathcal{B}$,

The set of all neighbourhoods of a given point in a topological space is a filter. Every filter base $\mathcal{B}$ generates a unique filter whose elements are the supersets of members of $\mathcal{B}$.

Let X be an inhabited set, and let U, V be subsets of the Cartesian product $X \times X$. We define certain associated subsets as follows:

$$U \circ V \equiv \{(x, y) : \exists_{z \in X} ((x, z) \in U \wedge (z, y) \in V)\},$$
$$U^1 \equiv U, \quad U^{n+1} \equiv U \circ U^n \quad (n = 1, 2, \ldots),$$
$$U^{-1} \equiv \{(x, y) : (y, x) \in U\},$$

and, for $x \in X$,
$$U[x] \equiv \{y \in X : (x,y) \in U\}.$$

We say that U is **symmetric** if $U = U^{-1}$.

Now let $\mathcal{U}$ be a family of subsets of $X \times X$. We say that $\mathcal{U}$ is a **quasi-uniform structure**, or a **quasi-uniformity**, on X if the following conditions hold:

U1 $\mathcal{U}$ is a filter on $X \times X$.

U2 For all $x,y \in X$, $x = y$ if and only $(x,y) \in U$ for each $U \in \mathcal{U}$.

U3 For each $U \in \mathcal{U}$ there exists $V \in \mathcal{U}$ such that $V^2 \subset U$.

U4 For each $U \in \mathcal{U}$ there exists $V \in \mathcal{U}$ such that $X \times X = U \cup \neg V$.

The elements of $\mathcal{U}$ are called the **entourages** of (the quasi-uniform structure on) X, and the pair $(X, \mathcal{U})$—or, loosely, X itself—is called a **quasi-uniform space**. We call $\mathcal{U}$ a **uniform structure**, or **uniformity**, and X a **uniform space**, if, in addition to **U1–U4**, for each $U \in \mathcal{U}$ we have $U^{-1} \in \mathcal{U}$; in that case, $U \cap U^{-1}$ is a symmetric entourage.

Axiom **U2** says that the intersection of all the entourages in $\mathcal{U}$ is the **diagonal**

$$\Delta \equiv \{(x,x) : x \in X\}$$

of $X \times X$. Note that, since the equality relation is symmetric, this axiom implies that if $x = y$, then $(y,x) \in U$ for each $U \in \mathcal{U}$.

Classically, property **U4** always holds with $V = U$. As will become clear in later work, it is important to postulate it in the constructive theory, since it is the only uniform-space axiom that provides us with alternatives that enable proof- and definition-by-cases.

We define the **standard (uniform) inequality** on a quasi-uniform space $(X, \mathcal{U})$ by

$$x \neq y \Leftrightarrow \exists_{U \in \mathcal{U}} ((x,y) \notin U) \vee \exists_{U \in \mathcal{U}} ((y,x) \notin U).$$

In view of **U2**, it is easily seen that $\neq$ is indeed an inequality relation on X. Moreover, if X is a uniform space, then $x \neq y$ if and only if there exists $U \in \mathcal{U}$ with $(x,y) \notin U$.

A quasi-uniform structure $\mathcal{U}$ on X induces a quasi-uniform structure $\mathcal{U}_Y$ on an inhabited subset Y of X: the entourages of $\mathcal{U}_Y$ are the sets $U \cap (Y \times Y)$ with $U \in \mathcal{U}$. Taken with the inequality and quasi-uniform structure induced by those on X, the set Y is called a **quasi-uniform subspace** of X. The quasi-uniform structure $\mathcal{U}_Y$ is called the **subspace quasi-uniform structure** on Y.

Every quasi-metric space (X, ρ) is a quasi-uniform space in which the quasi-uniformity consists of all supersets, in $X \times X$, of sets of the form

$$\{(x,y) : \rho(x,y) \leqslant \varepsilon\}$$

with $\varepsilon > 0$. It is straightforward to verify axioms **U1–U3**. To verify **U4**, we use the fact that for each $\varepsilon > 0$ and all $x, y \in X$, either $\rho(x,y) > \varepsilon/2$ or $\rho(x,y) < \varepsilon$. If ρ is a metric on X, then the corresponding quasi-uniformity is a uniformity.

Proposition 3.2.1 *Let U be an entourage of a quasi-uniform space $(X,\mathcal{U})$. Then for all $x, y \in X$, either $x \neq y$ or $(x, y) \in U$.*

Proof. By **U4**, there exists $V \in \mathcal{U}$ such that $X \times X = U \cup \neg V$. If $(x, y) \in \neg V$, then $x \neq y$, by our definition of the inequality on X. ∎

Corollary 3.2.2 *The inequality on a quasi-uniform space is tight.*

Proof. Let $(X,\mathcal{U})$ be a quasi-uniform space, and x, y points of X such that $\neg (x \neq y)$. Then by Proposition 3.2.1, $(x, y) \in U$ for each $U \in \mathcal{U}$; so, by axiom **U2**, $x = y$. ∎

Lemma 3.2.3 *Let V be an entourage of a quasi-uniform space, and n an integer $\geqslant 2$. Then V^n is an entourage, and $V^{n-1} \subset V^n$.*

Proof. Given $(x, y) \in V^{n-1}$, note, from axiom **U2**, that $(x, x) \in V$. Hence $(x, y) \in V \circ V^{n-1} = V^n$. In particular, $V \subset V^n$, so, by **U1** and **F2**, V^n is an entourage. ∎

Lemma 3.2.4 *Let $(X,\mathcal{U})$ be a quasi-uniform space, and $U \in \mathcal{U}$. Then there exists $V \in \mathcal{U}$ such that $V^3 \subset U$.*

Proof. Using **U3** twice, construct $W, V \in \mathcal{U}$ such that $W^2 \subset U$ and $V^2 \subset W$. By Lemma 3.2.3, $V \subset V^2 \subset W$, so $V^3 = V \circ V^2 \subset W \circ V^2 \subset W \circ W \subset U$. ∎

Lemma 3.2.5 *Let U be an entourage of a quasi-uniform space $(X,\mathcal{U})$. Then there exists an entourage V such that $V^2 \subset U$ and $\neg U \subset \sim V$.*

Proof. Using Lemma 3.2.4, choose an entourage V such that $V^3 \subset U$. We see from Lemma 3.2.3 that $V^2 \subset U$. By **U4**, there exists an entourage W such that $X \times X = V \cup \neg W$. Consider (x, y) in $\neg U$ and (s, t) in V. If $(x, s) \in V$ and $(t, y) \in V$, then $(x, y) \in V^3 \subset U$, a contradiction. Hence either $(x, s) \in \neg W$ and so $x \neq s$, or else $(t, y) \in \neg W$ and so $t \neq y$. Thus $(x, y) \neq (s, t)$. It follows that $\neg U \subset \sim V$. ∎

Proposition 3.2.6 *If U is an entourage of a quasi-uniform space $(X,\mathcal{U})$, then for each integer $n \geqslant 2$, there exists an entourage V such that $V^n \subset U$ and $X \times X = U \cup \sim V$.*

Proof. By **U4**, there exists an entourage W such that $X \times X = U \cup \neg W$. By the preceding lemma, there exists an entourage V such that $V^2 \subset W$ and $\neg W \subset \sim V$. The first of these properties implies that $V^2 \subset U$; the second, that $X = U \cup \sim V$. This completes the case $n = 2$. A simple induction argument now disposes of the case where n is a power of 2. In the general case, choose a positive integer k such that $2^k > n$. Then there exists $V \in \mathcal{U}$ such that $V^{2k} \subset U$ and $X \times X = U \cup {\sim}V$. By Lemma 3.2.3, $V^n \subset V^{2k}$, so $V^n \subset U$ and we are through. ∎

Corollary 3.2.7 *If U is an entourage of a uniform space $(X,\mathcal{U})$, then the conclusion of* Proposition 3.2.6 *holds for some symmetric entourage V.*

Proof. All we need do is replace V, as constructed in Proposition 3.2.6, by $V \cap V^{-1}$. ∎

For each positive integer n we define an n-**chain of entourages** of a quasi-uniform space $(X,\mathcal{U})$ to be an n-tuple $(U_1,\ldots,U_n)$ of entourages such that $U_k^2 \subset U_{k-1}$ and $X \times X = U_{k-1} \cup \neg U_k$ for each $k \geqslant 2$. Proposition 3.2.6 ensures that for each $U \in \mathcal{U}$ and each positive integer n there exists an n-chain $(U_1,\ldots,U_n)$ of entourages with $U_1 = U$; Corollary 3.2.7 shows that when $\mathcal{U}$ is a *uniform* structure, we can also ensure that the entourages U_k are symmetric for each $k \geqslant 2$.

We now prove the first two of a succession of little lemmas that will be useful on several occasions.

Lemma 3.2.8 *Let U,V be entourages of a quasi-uniform space $(X,\mathcal{U})$ such that $V^2 \subset U$, and let x,y be points of X such that $y \in V[x]$. Then $V[y] \subset U[x]$. If U and V are symmetric, then $V[x] \subset U[y]$.*

Proof. Given $z \in V[y]$, we have $(y,z) \in V$ and $(x,y) \in V$; whence $(x,z) \in V^2 \subset U$ and therefore $z \in U[x]$. Now suppose that U,V are symmetric. If $z \in V[x]$, then $(z,x) \in V^{-1} = V$ and $(x,y) \in V$, so $(z,y) \in V^2 \subset U$ and therefore, by symmetry, $z \in U[y]$. ∎

Lemma 3.2.9 *Let U be an entourage of a quasi-uniform space $(X,\mathcal{U})$, let $x \in X$, and let $S \subset X$. If $\{x\} \times S \subset {\sim}U$, then $U[x] \subset {\sim}S$. Let also $V \in \mathcal{U}$ be such that $V^2 \subset U$ and $X \times X = U \cup {\sim}V$. If $U[x] \subset {\sim}S$, then $\{x\} \times S \subset {\sim}V$.*

Proof. Assuming that $\{x\} \times S \subset {\sim}U$, let $y \in U[x]$ and $z \in S$. Then $(x,z) \in {\sim}U$ and $(x,y) \in U$, so $(x,z) \neq (x,y)$ in $X \times X$, and therefore $y \neq z$. This proves the first conclusion.

To prove the second, this time assume that $U[x] \subset {\sim}S$. For each $s \in S$ we have either $(x,s) \in U$ and therefore $s \in U[x] \subset {\sim}S$, which is absurd, or else, as must be the case, $(x,s) \in {\sim}V$. Hence $\{x\} \times S \subset {\sim}V$. ∎

A given quasi-uniform structure $\mathcal{U}$ on X gives rise to the corresponding **quasi-uniform topology** $\tau_{\mathcal{U}}$, in which the sets $U[x]$, with $U \in \mathcal{U}$, form a base of neighbourhoods of the point x. (Note that, by axiom **U2**, $x \in U[x]$ for each $U \in \mathcal{U}$.) This topology has the topological reverse Kolmogorov property: for if $S \in \tau_{\mathcal{U}}$ and $x \in S$, then we can find $U \in \mathcal{U}$ such that $U[x] \subset S$; if $y \notin S$, then $y \notin U[x]$, so $(x, y) \in \neg U$ and therefore $x \neq y$.

If $\mathcal{U}$ is a *uniform* structure, then the topology $\tau_{\mathcal{U}}$ is Hausdorff. To see this, consider x, y in X with $x \neq y$ in X. Pick U in $\mathcal{U}$ with $(x, y) \in \neg U$, and, using Corollary 3.2.7, construct entourages U, V such that V is symmetric, $V^3 \subset U$, and $X \times X = U \cup \sim V$. In view of the symmetry of the inequality relation, it suffices to show that $V[x] \subset \sim (V[y])$. To that end, let $s \in V[x]$ and $t \in V[y]$, and suppose that $(s, t) \in V$. Since also $(x, s) \in V$ and $(t, y) \in V^{-1} = V$, we have $(x, y) \in V^3$; so $(x, y) \in U$, a contradiction. It follows that $(s, t) \notin V$; whence $(s, t) \in \neg V$ and therefore $s \neq t$.

The quasi-uniform topology $\tau_{\mathcal{U}}$ gives rise to the corresponding topological point-set pre-apartness defined by

$$x \bowtie_{\tau_{\mathcal{U}}} S \Leftrightarrow \exists_{A \in \tau_{\mathcal{U}}} (x \in A \subset \sim S). \tag{3.15}$$

To show that this defines a point-set *apartness*, let $x \bowtie_{\tau_{\mathcal{U}}} S$; then there exists $U \in \mathcal{U}$ such that $U[x] \subset \sim S$. Pick $V \in \mathcal{U}$ such that $V^2 \subset U$. By Proposition 3.2.1, either $x \neq y$ or else $(x, y) \in V$. In the latter case, by Lemma 3.2.8, $y \in V[y] \subset U[x] \subset \sim S$; since $V[y]$ is a $\tau_{\mathcal{U}}$-neighbourhood of y, we therefore have $y \bowtie_{\tau_{\mathcal{U}}} S$.

We observed earlier that if $(X, \mathcal{U})$ is a uniform space, then the associated topological space $(X, \tau_{\mathcal{U}})$ is topologically Hausdorff; whence, by Proposition 2.4.5, the point-set pre-apartness $\bowtie_{\tau_{\mathcal{U}}}$ is Hausdorff.

Proposition 3.2.10 *Let $(X, \mathcal{U})$ be a quasi-uniform space. Then for all $x \in X$ and $S \subset X$,*

$$x \bowtie_{\tau_{\mathcal{U}}} S \Leftrightarrow \exists_{V \in \mathcal{U}} (\{x\} \times S \subset \sim V).$$

Proof. Suppose that $x \bowtie_{\tau_{\mathcal{U}}} S$. Then, by definition of $\tau_{\mathcal{U}}$, there exists $U \in \mathcal{U}$ such that $U[x] \subset \sim S$. Choosing $V \in \mathcal{U}$ such that $V^2 \subset U$ and $X \times X = U \cup \sim V$, we have $\{x\} \times S \subset \sim V$, by the last part of Lemma 3.2.9.

Conversely, suppose that $\{x\} \times S \subset \sim V$ for some $V \in \mathcal{U}$. By the first part of Lemma 3.2.9, $V[x] \subset \sim S$; since $V[x]$ is a $\tau_{\mathcal{U}}$-neighbourhood of x, it follows that $x \bowtie_{\tau_{\mathcal{U}}} S$. ∎

Given a quasi-uniform space $(X, \mathcal{U})$, we define the corresponding **quasi-uniform** (or, if $\mathcal{U}$ is a uniform structure, **uniform**) **pre-apartness** by

$$S \bowtie_{\mathcal{U}} T \Leftrightarrow \exists_{U \in \mathcal{U}} (S \times T \subset \sim U). \tag{3.16}$$

We know from Proposition 3.2.10 and the remarks preceding it that the corresponding point-set relation, in which S is a singleton, is just the point-set apartness $\bowtie_{\tau_{\mathcal{U}}}$ induced on X by the quasi-uniform topology $\tau_{\mathcal{U}}$. We denote by $-S$ the apartness complement of S relative to this point-set apartness. We aim to show that (3.16) really does define a set-set pre-apartness. First, though, we show that we can replace the complement on the right of (3.16) by the logical complement.

Proposition 3.2.11 *Let* S, T *be subsets of a quasi-uniform space* $(X, \mathcal{U})$. *Then* $S \bowtie_{\mathcal{U}} T$ *if and only if there exists* $V \in \mathcal{U}$ *such that* $S \times T \subset \neg V$.

Proof. "Only if" is clear, since $\sim U \subset \neg U$ for any $U \in \mathcal{U}$. To prove "if", suppose there exists $V \in \mathcal{U}$ such that $S \times T \subset \neg V$. Using Proposition 3.2.6, pick $W \in \mathcal{U}$ such that $W^2 \subset V$ and $X \times X = V \cup \sim W$. Then $S \times T \subset \sim W$ and therefore $S \bowtie_{\mathcal{U}} T$. ∎

In places we come across terms like $-U[x], \sim U[x]$, and $\neg U[x]$, where $U \subset X \times X$. Since we have not yet dealt with products of uniform spaces and their resulting pre-apartness relations, there should be little doubt that $-U[x]$ is meant to be interpreted as $-(U[x])$. It is straightforward to show that

$$(\neg U)[x] = \neg (U[x]),$$

so it makes no difference which of the two interpretations we give to $\neg U[x]$. However, as we now digress to show, with the interpretation of $\sim U[x]$ things are different. Note that when we apply concepts like *open*, *closed*, *interior*, and *closure* to an entourage of X, we are referring to the product topology derived from the quasi-uniform topology on X.

Lemma 3.2.12 *Let* X *be a set with an inequality,* U *a subset of* $X \times X$, *and* x *an element of* X. *Then*

$$(\sim U)[x] \subset \sim (U[x]). \tag{3.17}$$

If $(X, \mathcal{U})$ *is a quasi-uniform space and* U *is an open entourage of* X, *then*

$$(\sim U)[x] = \sim (U[x]).$$

Proof. If $y \in (\sim U)[x]$, then $(x, y) \in \sim U$, so for each $z \in U[x]$ we have $(x, y) \neq (x, z)$ and therefore $y \neq z$. Hence (3.17) holds.

Now assume that $(X, \mathcal{U})$ is a quasi-uniform space and that U is an open entourage. Let $y \in \sim (U[x])$ and consider any $(s, t) \in U$. Since U is open, the definition of the product topology on $X \times X$ tells us that there exist $V_1, V_2 \in \mathcal{U}$ such that $(s, t) \in V_1[s] \times V_2[t] \subset U$. Setting $V \equiv V_1 \cap V_2$, we have $V \in \mathcal{U}$ (by **U1** and **F1**) and $V[s] \times V[t] \subset U$. Now choose $W \in \mathcal{U}$ such that (V, W) is a 2-chain. If both $(s, x) \in V$ and $(t, y) \in V$, then $(x, y) \in V[s] \times V[t]$, so $(x, y) \in U$ and therefore $y \in U[x]$, a contradiction. Hence either $(s, x) \in \neg W$ and so $s \neq x$, or $(t, y) \in \neg W$ and so $t \neq y$; in each case we have $(s, t) \neq (x, y)$. Thus $(x, y) \in \sim U$ and so $y \in (\sim U)[x]$. From this and (3.17) we conclude that $(\sim U)[x] = \sim (U[x])$. ∎

Next we show that the (classically valid) opposite inclusion to that in (3.17) cannot be proved in general.

Proposition 3.2.13 *The following are equivalent:*

(i) **MP**$_{or}$ *(see page 59).*

(ii) *Let* $\mathbf{R}_d$ *denote the space of real numbers taken with the denial inequality, and let* $U \equiv \{(0,0)\}$. *Then*

$$\sim (U[x]) = (\sim U)\,[x]$$

for all $x \in \mathbf{R}_d$, *where on the left (respectively, right)* $\sim$ *denotes the complement with respect to the denial inequality on* $\mathbf{R}_d$ *(respectively, the inequality on the product space* $\mathbf{R}_d \times \mathbf{R}_d$).

Proof. First suppose that **MP**$_{or}$ holds, and let x_1, x_2 be any points of $\mathbf{R}_d$ such that $x_2 \in \sim (U[x_1]) = \neg (U[x_1])$, where $U = \{(0,0)\}$. Applying Lemma 2.5.17, we find that either $\neg (x_1 = 0)$ or $\neg (x_2 = 0)$; whence $(x_1, x_2) \neq (0,0)$ in the product space $\mathbf{R}_d \times \mathbf{R}_d$, and therefore $x_2 \in (\sim U)\,[x_1]$. Thus $\sim (U[x_1]) \subset (\sim U)\,[x_1]$. Reference to (3.17) completes the proof that (i) implies (ii).

Conversely, assuming (ii), let x_1, x_2 be real numbers such that

$$\neg (x_1 = 0 \wedge x_2 = 0)\,.$$

If $z \in U[x_1]$, then $(x_1, z) = (0,0)$, so $x_1 = 0 = z$; hence $\neg (x_2 = 0)$ and therefore $\neg (x_2 = z)$. Thus $x_2 \in \neg (U\,[x_1]) = \sim (U[x_1])$; whence, by (ii), $x_2 \in (\sim U)\,[x_1]$. Then $(x_1, x_2) \in \sim U$, so $(x_1, x_2) \neq (0,0)$ and therefore, by definition of the inequality on the product $\mathbf{R}_d \times \mathbf{R}_d$, either $\neg (x_1 = 0)$ or $\neg (x_2 = 0)$. It follows from this and Lemma 2.5.17 that (ii) implies (i). ∎

Bearing in mind the preceding two results, we adopt the convention that

$$\sim U[x] \text{ stands for } \sim (U[x]).$$

Returning now to the main stream of the chapter, in order to examine various types of entourage, we next prove

Lemma 3.2.14 *Let* X *be a topological space. Then for each open (respectively, closed) subset* U *of* $X \times X$, *and each* x *in* X, *the set* $U[x]$ *is open (respectively, closed) in* X.

Proof. The mapping $\theta : y \rightsquigarrow (x,y)$ of X onto $\{x\} \times X$ is topologically continuous (this is left as an exercise). If U is open in $X \times X$, then $U \cap (\{x\} \times X)$ is open in the subspace $\{x\} \times X$ of $X \times X$, so

$$U[x] = \theta^{-1}\,(U \cap (\{x\} \times X))$$

is open in X. A similar argument applies when U is closed in $X \times X$. ∎

Lemma 3.2.15 *Let* U *be a symmetric entourage of a uniform space* $(X, \mathcal{U})$. *Then the interior (in the product topology on* $X \times X$) *of* U^3 *is an entourage of* X.

Proof. If $(x, y) \in U$, then $U[x] \times U[y]$ is a neighbourhood of (x, y) in the product topological space $X \times X$ (by definition of $\tau_{\mathcal{U}}$), and

$$U[x] \times U[y] \subset U^{-1} \circ U \circ U = U^3.$$

Hence U is a subset of the interior of U^3. Since $\mathcal{U}$ is a filter on $X \times X$, the result follows. ∎

Proposition 3.2.16 *If U is an entourage of a uniform space $(X, \mathcal{U})$, then so is the interior of U.*

Proof. By Corollary 3.2.7, there exists a symmetric $V \in \mathcal{U}$ such that $V^3 \subset U$. Lemma 3.2.15 shows that $(V^3)^{\circ} \in \mathcal{U}$. Since $(V^3)^{\circ} \subset U^{\circ}$ and $\mathcal{U}$ is a filter, the desired conclusion follows. ∎

Lemma 3.2.17 *Let U be an entourage of a uniform space $(X, \mathcal{U})$. Then there exists a symmetric open entourage V of X whose closure is a subset of U.*

Proof. Using Corollary 3.2.7, construct a symmetric entourage W such that $W^5 \subset U$. If (x, y) is in the closure of W^3, then, since $W[x] \times W[y]$ is a neighbourhood of (x, y) in the product topology on $X \times X$, the set $W^3 \cap (W[x] \times W[y])$ is inhabited. It follows that

$$(x, y) \in W \circ W^3 \circ W^{-1} = W^5 \subset U.$$

In view of this and Lemma 3.2.15, it suffices to take V equal to the interior of W^3 in the product topology. ∎

Let $(X, \mathcal{U})$ be a quasi-uniform space. A subset $\mathcal{B}$ of $\mathcal{U}$ is called a **base of entourages** if for each $U \in \mathcal{U}$, there exists $B \in \mathcal{B}$ with $B \subset U$.

Proposition 3.2.18 *The symmetric open entourages form a base of entourages of a uniform space.*

Proof. This is a simple consequence of Proposition 3.2.16. ∎

Proposition 3.2.19 *Let $(X, \mathcal{U})$ be a quasi-uniform space with a base of open entourages. Then the topological space $(X, \tau_{\mathcal{U}})$ is topologically locally decomposable and topologically consistent.*

Proof. Let $A \in \tau_{\mathcal{U}}$ and $x \in A$. By definition of $\tau_{\mathcal{U}}$, there exists $U \in \mathcal{U}$ such that $x \in U[x] \subset A$. By Proposition 3.2.6 and our hypotheses, there exists an open entourage V such that $V^2 \subset U$ and $X \times X = U \cup \sim V$. Then $x \in V[x]$ and, by Lemma 3.2.14, $V[x]$ is in $\tau_{\mathcal{U}}$. Moreover, for each $y \in X$ we have either $y \in U[x]$ or else $y \in \sim V[x]$ (see Lemma 3.2.12); so $X = A \cup \sim V[x]$. This completes the proof of topological local decomposability, from which, via Proposition 2.2.12, we see that $(X, \tau_{\mathcal{U}})$ is topologically consistent. ∎

Corollary 3.2.20 *On a uniform space $(X, \mathcal{U})$, the uniform topology and the apartness topology induced by $\bowtie_{\mathcal{U}}$ coincide.*

Proof. By Proposition 3.2.18, the open entourages form a base of entourages for X. Hence $(X, \tau_{\mathcal{U}})$ is topologically consistent, by Proposition 3.2.19. Since, by Proposition 3.2.10, the apartness induced by $\mathcal{U}$ coincides with that induced by $\tau_{\mathcal{U}}$, the desired result follows. ∎

The next three lemmas will enable us to establish that a quasi-uniform pre-apartness is actually a set-set apartness.

Lemma 3.2.21 *If (U, V) is a 2-chain of entourages in a quasi-uniform space X, then $\neg\neg V[x] \subset U[x]$ for each $x \in X$.*

Proof. Let $y \in \neg\neg V[x]$. Either $(x, y) \in U$ or $(x, y) \in \neg V$. The latter case is ruled out, since it entails the contradiction $y \in \neg V[x]$. ∎

Lemma 3.2.22 *Let (U, V) be a 2-chain of entourages of a quasi-uniform space, let $x \in X$, and let $S \subset X$. If $U[x] \subset \neg S$, then $V[x] \subset -S$.*

Proof. Suppose that $U[x] \subset \neg S$. For any $y \in V[x]$, Lemma 3.2.8 shows that $V[y] \subset U[x] \subset \neg S$. We therefore have $\mathbf{apart}(y, S)$, by Proposition 3.2.11. It follows from this and Proposition 3.2.10 that $V[x] \subset -S$. ∎

Lemma 3.2.23 *If (U, V) is a 2-chain of entourages of a quasi-uniform space, then $V[x] \subset -\neg U[x]$ for each $x \in X$.*

Proof. Since $U[x] \subset \neg\neg U[x]$, the result follows from Lemma 3.2.22. ∎

From now on, as in the following proof, we shall use Proposition 3.2.11 without explicitly referring to it.

Proposition 3.2.24 *The relation $\bowtie_{\mathcal{U}}$, defined at (3.16), between subsets of a quasi-uniform space $(X, \mathcal{U})$ is an apartness relation that has the Efremovič property and satisfies $\mathbf{A4_{ss}}$. If $\mathcal{U}$ is a uniformity, then $\bowtie_{\mathcal{U}}$ is symmetric.*

Proof. Since $X \times \varnothing = \varnothing \subset \sim(X \times X)$ and $X \times X \in \mathcal{U}$, axiom **B1** holds. Since, by Proposition 3.2.10, the point-set version of $\bowtie_{\mathcal{U}}$ coincides with the point-set apartness $\bowtie_{\tau_{\mathcal{U}}}$ induced by the uniform topology $\tau_{\mathcal{U}}$, we see that $\bowtie_{\mathcal{U}}$ satisfies both **B2** and **B4**.

To deal with **B3**, let $A \bowtie_{\mathcal{U}} (B_1 \cup B_2)$ and choose $U \in \mathcal{U}$ such that $A \times (B_1 \cup B_2) \subset \neg U$; then $A \times B_1 \subset \neg U$ and $A \times B_2 \subset \neg U$, so $A \bowtie_{\mathcal{U}} B_1$ and $A \bowtie_{\mathcal{U}} B_2$. Now suppose, conversely, that $A \bowtie_{\mathcal{U}} B_1$ and $A \bowtie_{\mathcal{U}} B_2$. Then there exist $U, V \in \mathcal{U}$ such that $A \times B_1 \subset \neg U$ and $A \times B_2 \subset \neg V$; whence $A \times (B_1 \cup B_2) \subset \neg(U \cap V)$. But **U1** and **F1** together tell us that $U \cap V \in \mathcal{U}$, so $A \bowtie_{\mathcal{U}} (B_1 \cup B_2)$. Hence

$$A \bowtie_{\mathcal{U}} (B_1 \cup B_2) \Leftrightarrow A \bowtie_{\mathcal{U}} B_1 \wedge A \bowtie_{\mathcal{U}} B_2. \tag{3.18}$$

A similar argument shows that

$$(A_1 \cup A_2) \bowtie_{\mathcal{U}} B \Leftrightarrow A_1 \bowtie_{\mathcal{U}} B \wedge A_2 \bowtie_{\mathcal{U}} B,$$

from which, together with (3.18), we readily derive the conclusion of **B3**.

To handle **B5**, let $x \in -S$, let (U_1, U_2, U_3) be a 3-chain of entourages such that $\{x\} \times S \subset \neg U_1$, and let $T \equiv \neg U_3[x]$. Then $x \in U_3[x] \subset \neg T$, so $\{x\} \times T \subset \neg U_3$; whence $x \in -T$. On the other hand, for each $y \in X$ either $(x,y) \in \neg U_3$ or $(x,y) \in U_2$. In the first case, $y \in \neg U_3[x] = T$. In the second case, $y \in U_2[x]$; since $U_1[x] \subset \neg S$, it follows from Lemma 3.2.22 that $y \in -S$. Hence $X = -S \cup T$.

Now for the Efremovič property. Let $S \bowtie_{\mathcal{U}} T$ in X, and construct a 2-chain (U,V) of entourages such that $S \times T \subset \neg U$. Let

$$ E \equiv \{x \in X : \exists_{s \in S} ((s,x) \in V)\} . $$

Then $S \times \neg E \subset \neg V$ and therefore $S \bowtie_{\mathcal{U}} \neg E$. On the other hand, if $t \in T$, $x \in E$, and $(x,t) \in U_2$, then there exists $s \in S$ such that $(s,t) \in U_2 \circ U_2 \subset U_1$, which is absurd. Hence $E \times T \subset \neg U_2$ and therefore $E \bowtie_{\mathcal{U}} T$. This proves that **EF** holds in X; whence, by Proposition 3.1.10, **A4**$_{ss}$ also holds. Finally, it is clear that $\bowtie_{\mathcal{U}}$ is symmetric when $\mathcal{U}$ is a uniformity. ∎

By a **(quasi-)uniform apartness space** we mean a (quasi-)uniform space taken with its (quasi-)uniform apartness $\bowtie_{\mathcal{U}}$. Note that this apartness is $\mathbf{T}_1$: for if $x \neq y$ in our quasi-uniform space, then we may assume that there exists $U \in \mathcal{U}$ such that $(x,y) \notin U$; whence $\{x\} \times \{y\} \subset \neg U$ and therefore, by Proposition 3.2.11, $x \bowtie_{\mathcal{U}} \{y\}$.

Proposition 3.2.25 *Let $(X, \mathcal{U})$ be a quasi-uniform space, Y an inhabited subset of X, and $\mathcal{U}_Y$ the quasi-uniform structure induced on Y by $\mathcal{U}$. Then*

$$ \forall_{S,T \subset Y} (S \bowtie_{\mathcal{U}_Y} T \Leftrightarrow S \bowtie_{\mathcal{U}} T) . $$

Proof. This follows from the definition of $\mathcal{U}_Y$ and the fact that

$$ S \times T \subset X \cap \neg U \Leftrightarrow S \times T \subset Y \cap \neg (U \cap (Y \times Y)) $$

whenever $S, T \subset Y$ and $U \in \mathcal{U}$. ∎

We remarked on page 23 that every inhabited subset of a point-set apartness space is a point-set apartness subspace. In the case of quasi-uniform spaces, this result can also be proved from Proposition 3.2.25, with reference to the definition of the quasi-uniform apartness on a subset and the definition of *apartness subspace*. Proposition 3.2.25 actually gives us a little more: namely, that the subspace apartness coincides with the quasi-uniform apartness on the quasi-uniform subspace.

Quasi-uniform apartness spaces have many interesting properties, some of which, like the one we now define, would not be significant in a classical development:

B7 $\forall_{S,T \subset X} [S \bowtie T \Rightarrow \forall_{x \in X} \exists_{R \subset X} (x \in -R \wedge (S - R \neq \emptyset \Rightarrow \neg R \bowtie T))]$.

This property was adopted as an axiom in an earlier version of our theory of set-set apartness [25]. It says that if we start with two sets S,T that are apart, then each $x \in X$ has a neighbourhood $-R$ in the apartness topology such that if that neighbourhood intersects S, then x is so close to S that not only $-R$, but also the larger neighbourhood $\neg R$, is apart from T. It is almost trivial that if a pre-apartness space has the property **B7**, then it has the weak nested neighbourhoods property.

Proposition 3.2.26 *A uniform apartness space satisfies* **B7**.

Proof. Let $S \bowtie T$ in the uniform apartness space $(X, \mathcal{U})$, and let $x \in X$. Corollary 3.2.7 enables us to construct a 3-chain (U, V, W) such that $S \times T \subset \neg U$, V is symmetric, and $V^3 \subset U$. Setting $R \equiv \neg W[x]$, we have $\{x\} \times R \subset \neg W$ and therefore $x \in -R$ (by Proposition 3.2.11). Also, $\neg R = \neg\neg W[x] \subset V[x]$. Suppose there exists $s \in S - R$; then $s \in \neg R$, so $(x, s) \in V$ and therefore $(s, x) \in V$. Likewise, for any $y \in \neg R$ we have $(x, y) \in V$. Given $t \in T$, assume that $(y, t) \in V$. Then $(s, t) \in V \circ V \circ V \subset U$; this is absurd, since $\{s\} \times T \subset S \times T \subset \neg U$. We conclude that $(y, t) \notin V$. Hence $\neg R \times T \subset \neg V$ and therefore, by Proposition 3.2.11, $\neg R \bowtie T$. ∎

Taking Proposition 3.2.26 together with the remark immediately preceding it, we obtain

Corollary 3.2.27 *A uniform apartness space has the weak nested neighbourhoods property.*

We say that a quasi-uniformity $\mathcal{U}$ on an apartness space $(X, \bowtie)$ is **compatible** with, or **induces**, the given apartness if the apartness corresponding to $\mathcal{U}$ is the same as $\bowtie$, a condition which (by Proposition 3.2.11) is equivalent to

$$S \bowtie T \Leftrightarrow \exists_{U \in \mathcal{U}} (S \times T \subset \neg U).$$

In the classical theory of proximity spaces, every symmetric proximity space with the Efremovič property has uniform structures that are compatible with the denial apartness (the one in which two sets are apart if and only if they are not near each other); see [75], page 71. The next proposition will enable us to show that there is no hope of proving constructively that the apartness obtained from the standard metric point-set apartness on **R** by the construction in Proposition 3.1.20 has any compatible quasi-uniformities.

Before stating the proposition, we comment that, in contrast to the classical situation, it is not provable constructively that the only subsets of a singleton set are the set itself and the empty set: for, given any singleton $\{x\}$ and any statement P, we can form the subset

$$S \equiv \{y \in \{x\} : P\},$$

and we are not entitled to claim that

$$S = \varnothing \vee S = \{x\}$$

unless we can prove $P \vee \neg P$.

Proposition 3.2.28 *If there exists a pre-apartness space X such that*

(i) $A \bowtie B \Rightarrow \forall_{x \in X} (x \notin A \vee x \notin B)$ *and*

(ii) *any two disjoint subsets of a singleton in X are apart (that is, if $x \in X$, $A \subset \{x\}$, $B \subset \{x\}$, and $A \cap B = \varnothing$, then $A \bowtie B$),*

then the weak law of excluded middle

$$\neg P \vee \neg\neg P$$

holds.

Proof. Suppose there exists such a pre-apartness space X. Fixing $\xi \in X$, consider any statement P, and define

$$
\begin{aligned}
A &\equiv \{x \in X : (x = \xi) \wedge P\}, \\
B &\equiv \{x \in X : (x = \xi) \wedge \neg P\}.
\end{aligned}
$$

Then A and B are disjoint subsets of $\{\xi\}$, so, by hypothesis (ii), $A \bowtie B$. By hypothesis (i), either $\xi \notin A$, in which case $\neg P$ holds, or else $\xi \notin B$ and therefore $\neg\neg P$ holds. ∎

   The somewhat eccentric hypothesis (ii) in the preceding proposition holds classically for any pre-apartness space X containing at least two distinct points. For in that case, if A, B are disjoint subsets of a singleton $\{x\}$ in X, then either $A = B = \varnothing$ and Lemma 3.1.2 applies; or else one of A, B is $\{x\}$ and the other is empty, when axiom **B1**, Lemma 3.1.1, and Proposition 3.1.4 give $A \bowtie B$.
   The hypothesis (ii) holds constructively if the apartness on X is induced by a quasi-uniform structure $\mathcal{U}$: for if $x \in X$, and A, B are disjoint subsets of $\{x\}$, then, taking any $U \in \mathcal{U}$, we have $A \times B = \varnothing \subset \neg U$; whence $A \bowtie B$.
   For convenience, if X is a quasi-metric space, $x \in X$, and $S \subset X$, we shall write $\rho(x, S) > 0$ to signify that

$$\exists_{r > 0} \forall_{y \in S} (\rho(x, y) > r)$$

even when S is not known to be located.

Corollary 3.2.29 *Let ρ denote the standard metric on $\mathbf{R}$, let X be an inhabited subset of $\mathbf{R}$, and let $\bowtie_0$ be the symmetric set-set pre-apartness defined on X by*

$$A \bowtie_0 B \Leftrightarrow \forall_{x \in X} (\rho(x, A) > 0 \vee \rho(x, B) > 0).$$

If $\bowtie_0$ is induced by a uniform structure, then the weak law of excluded middle holds.

Proof. Proposition 3.1.20 tells us that $\bowtie_0$ is a pre-apartness on X. It is clear that $\bowtie_0$ satisfies condition (i) of the preceding proposition. If it is induced by a uniform structure, then, as the remark preceding this corollary explains, hypothesis (ii) of Proposition 3.2.28 holds. The result now follows from that proposition. ∎

Classically, the (proximity corresponding to the) pre-apartness defined in the preceding corollary has the Efremovič property, and so has compatible uniformities. The corollary shows that, in a sense, the constructive theory of apartness spaces is strictly bigger than that of uniform spaces. It also has this interesting consequence: the compact metric subspace $[0,1]$ of $\mathbf{R}$ has a pre-apartness—the one induced by the set-set pre-apartness in Corollary 3.2.29—which induces the original metric topology on $[0,1]$ but which we cannot prove to be the same as the metric apartness. (This contrasts with the classical situation in which, for each compact Hausdorff topological space, there exists a unique proximity structure whose apartness topology coincides with the original compact topology; see Theorem 41.1 of [93].) To see this, take $X \equiv [0,1]$ in Corollary 3.2.29. Let $x \in X$, $A \subset X$, and $x \bowtie_0 A$. Then since $\rho(x,x) = 0$, we have $\rho(x,A) > 0$, so we can pick $r > 0$ such that $|x - y| > 2r$ for each $y \in A$. Then $B(x,r) \cap [0,1]$ is a subset of the apartness complement $-_0 A$ of A relative to $\bowtie_0$. It follows that every $\bowtie_0$-open neighbourhood of x in $[0,1]$ contains a neighbourhood of x in the metric topology on $[0,1]$. Conversely, if, for each $r > 0$, we define

$$ S \equiv \left\{ y \in [0,1] : |x - y| \geqslant \frac{r}{2} \right\}, $$

then we have $x \in -_0 S \subset B(x,r)$. Hence the apartness topology induced by $\bowtie_0$ coincides with the standard metric topology on $[0,1]$. However, if $\bowtie_0$ were provably the same as the metric apartness on $[0,1]$, then, according to Corollary 3.2.29, we could derive the weak law of excluded middle.

Corollary 3.2.30 *Let X be an inhabited set with the denial inequality. Then*

$$ A \bowtie_\varnothing B \Leftrightarrow (A = \varnothing \vee B = \varnothing) \tag{3.19} $$

defines a symmetric set-set apartness on X that satisfies the Efremovič condition. If this apartness is induced by a uniform structure, then the weak law of excluded middle holds.

Proof. It is routine to verify that the point-set relation $\bowtie_\varnothing$ defined by

$$ x \bowtie_\varnothing B \Leftrightarrow B = \varnothing $$

satisfies **A1–A5**. Moreover, it is locally decomposable, since if $x \bowtie_\varnothing U$ and therefore $U = \varnothing$, we have $X = -U \cup U$. Noting that

$$ \forall_{x \in X} (x \bowtie_\varnothing A \vee x \bowtie_\varnothing B) \Leftrightarrow (A = \varnothing \vee B = \varnothing), $$

we now see from Proposition 3.1.20 that the set-set relation defined at (3.19) is a symmetric set-set apartness. As in the proof of Corollary 3.2.29, we also see that if it is induced by a uniform structure, then the weak law of excluded middle holds. Finally, consider A,B with $A \bowtie_\varnothing B$. If $A = \varnothing$, then $A \bowtie_\varnothing \neg\varnothing$ and $\varnothing \bowtie_\varnothing B$. If $B = \varnothing$, then $A \bowtie_\varnothing \neg X$ and $X \bowtie_\varnothing B$. Thus $\bowtie_\varnothing$ satisfies EF. ∎

This corollary shows even more clearly than does Corollary 3.2.29 that our theory of apartness spaces is larger than that of uniform spaces. For it provides an example of a symmetric set-set apartness for which, without the weak law of excluded middle, we cannot produce an inducing uniform structure *even though the apartness satisfies the Efremovič condition.*

We emphasise that the set-set pre-apartness relations in the preceding two corollaries should not be regarded as pathological: for example, the first can be described classically by saying that $A \bowtie B$ if and only if the closures of A and B in the metric topology on $\mathbf{R}$ are disjoint; as we mentioned on page 73, the negation of this pre-apartness is a standard classical example of a proximity.

We now return briefly to a topological group (G, τ) and its associated left pre-apartness $\bowtie^{(L)}$. Retaining the notation used in the work on topological groups at the end of the previous section, for each $A \in \mathfrak{N}(e)$ we define

$$U_A \equiv \left\{ (x,y) \in G \times G : x^{-1}y \in A \right\}.$$

The sets U_A, with $A \in \mathfrak{N}(e)$, form a filter base on $G \times G$. After some preliminaries, we shall show that under certain reasonable conditions on G, the filter $\mathcal{U}$ generated by the sets U_A is a uniform structure, the **left uniform structure**, and that the associated uniform topology and apartness are precisely the original ones on the topological group G.

Lemma 3.2.31 *Let G be a topological group with the topological reverse Kolmogorov property, and let $x, y \in G$.*

(i) *If there exists $A \in \mathfrak{N}(e)$ such that $(x,y) \in \neg U_A$, then $x \neq y$.*

(ii) *If the inequality on G is tight and $(x,y) \in U_A$ for each $A \in \mathfrak{N}(e)$, then $x = y$.*

Proof. If $(x,y) \notin U_A$, then $x^{-1}y \notin A$, so $y \notin xA$. Since $xA \in \mathfrak{N}(x)$, it follows from the topological reverse Kolmogorov property that $y \neq x$; whence $x \neq y$. On the other hand, under the hypotheses of (ii), it follows from (i) that $\neg(x \neq y)$; whence, by the tightness of the inequality, $x = y$. ∎

Lemma 3.2.32 *Let (G, τ) be a topological group, and let A, B be neighbourhoods of e such that $BB \subset A$ and $G = A \cup {\sim}B$. Then $U_B \circ U_B \subset U_A$ and $G \times G = U_A \cup {\sim}U_B$.*

Proof. Let $(x, y) \in U_B$ and $(y, z) \in U_B$. Then $x^{-1}y \in B$ and $y^{-1}z \in B$, so

$$x^{-1}z = (x^{-1}y)(y^{-1}z) \in BB \subset A$$

and therefore $(x, z) \in U_A$. This establishes that $U_B \circ U_B \subset U_A$.

For the second conclusion of the lemma, consider any $x, y \in G$. Either $x^{-1}y \in A$ and therefore $(x, y) \in U_A$, or else $x^{-1}y \in {\sim}B$. In the latter event, for each $(s, t) \in U_B$ we have $x^{-1}y \neq s^{-1}t$; so either $x \neq s$ or $y \neq t$, and therefore $(x, y) \neq (s, t)$. Hence $(x, y) \in {\sim}U_B$. ∎

Lemma 3.2.33 *Let (G, τ) be a decomposable topological group that has the reverse Kolmogorov property and a tight inequality, and let $x, y \in G$. Then $x = y$ if and only if $(x, y) \in U_A$ for each $A \in \mathfrak{N}(e)$.*

Proof. Given $A \in \mathfrak{N}(e)$ and using Lemma 3.1.23, construct $B \in \mathfrak{N}(e)$ such that $BB \subset A$ and $G = A \cup {\sim}B$. If $(x, y) \in {\sim}U_B$, then, by Lemma 3.2.31, $x \neq y$. It follows that if $x = y$, then $(x, y) \in U_A$. This proves "only if". Reference to Lemma 3.2.31(ii) yields "if". ∎

By a $\mathbf{T}_1$ **topological space** we mean a topological space (X, τ) with the property:

$$x \neq y \Rightarrow \exists_{U \in \tau} (x \in \neg U \wedge y \in U).$$

Lemma 3.2.34 *The following conditions on a topological group G are equivalent:*

(i) *G is a $\mathbf{T}_1$ topological group.*

(ii) *For each $x \in G$ with $x \neq e$, there exists $A \in \mathfrak{N}(e)$ such that $x \notin A$.*

(iii) *For all $x, y \in G$ with $x \neq y$, there exists $A \in \mathfrak{N}(e)$ such that $(x, y) \in \neg U_A$.*

Proof. Clearly, (i) implies (ii). Assuming (ii), let $x \neq y$ in G. Then $y^{-1}x \neq e$, so there exists $A \in \mathfrak{N}(e)$ such that $y^{-1}x \notin A$ and therefore $x \notin yA$. Since $yA \in \mathfrak{N}(y)$, we conclude that (i) holds. Moreover, $x^{-1}y \notin A^{-1}$, so $(x, y) \in \neg U_{A^{-1}}$, where $A^{-1} \in \mathfrak{N}(e)$. Hence (ii) implies (iii). Finally, taking $y = e$ in (iii), we readily obtain (ii). ∎

These lemmas bring us back to the uniform structure on a topological group, for which we recall that in the presence of decomposability, the left pre-apartness is actually an apartness.

Proposition 3.2.35 *Let (G, τ) be a decomposable topological group that has the reverse Kolmogorov property and a tight inequality. Then the sets U_A, with $A \in \mathfrak{N}(e)$, form a filter base that generates a uniform structure $\mathcal{U}_G$ on G. Moreover, the topologies $\tau, \tau_{\mathcal{U}_G}$ coincide, and the uniform apartness $\bowtie_{\mathcal{U}_G}$ on G coincides with the left apartness $\bowtie^{(L)}$ on G. If also G is $\mathbf{T}_1$, then the inequality on G coincides with the uniform inequality associated with $\mathcal{U}_G$.*

Proof. We already know that $\mathcal{U}_G$ is a filter and, in view of Lemma 3.2.33, that it satisfies **U2**. To verify **U3** and **U4**, given $U \in \mathcal{U}_G$, we first pick $A \in \mathfrak{N}(e)$ such that $U_A \subset U$, then use Lemma 3.1.23 to construct $B \in \mathfrak{N}(e)$ such that $BB \subset A$ and $G \times G = A \cup {\sim}B$, and finally apply Lemma 3.2.32. Moreover, $B^{-1} \in \mathfrak{N}(e)$ and $U_{B^{-1}} \subset U_{A^{-1}} = U_A^{-1} \subset U^{-1}$, so $U^{-1} \in \mathcal{U}$. Hence $\mathcal{U}_G$ is symmetric and therefore a uniform structure on G.

Now, $xA = U_A[x]$ for each $x \in G$ and each $A \in \mathfrak{N}(e)$. Since, for such A, the sets xA form a base of neighbourhoods of x in the topology τ, and the sets $U_A[x]$ form a base of neighbourhoods of x for the uniform topology $\tau_{\mathcal{U}}$, these two topologies on G coincide. The coincidence of the apartness relations $\bowtie_{\mathcal{U}}$ and $\bowtie^{(L)}$ follows from the observation that, for subsets S, T of G and $A \in \mathfrak{N}(e)$, $S \times T \subset {\sim}U_A$ if and only if $S^{-1}T \subset {\sim}A$. Finally, if also G is $\mathbf{T}_1$, then it follows from Lemma 3.2.34 that the original inequality on G coincides with the uniform inequality associated with $\mathcal{U}_G$. ∎

3.3 Strong and Uniform Continuity

A mapping $f : X \to Y$ between set-set pre-apartness spaces is said to be **strongly continuous** if for all subsets S, T of X,

$$f(S) \bowtie f(T) \Rightarrow S \bowtie T.$$

It should be clear that

▷ according to the definitions in Section 2.3, when we regard X and Y as point-set pre-apartness spaces, such a mapping is continuous;

▷ if also Y is a $\mathbf{T}_1$-space, then the mapping is strongly extensional; and

▷ the composition of two strongly continuous mappings is strongly continuous.

If $(X, \mathcal{U}_X)$ and $(Y, \mathcal{U}_Y)$ are quasi-uniform spaces, then there is another natural notion of continuity that can be applied to a mapping $f : X \to Y$: we say that f is **uniformly continuous** if for each $V \in \mathcal{U}_Y$ there exists $U \in \mathcal{U}_X$ such that $(f \times f)(U) \subset V$, where

$$(f \times f)(x, y) \equiv (f(x), f(y)) \qquad (x, y \in X).$$

The composition of two uniformly continuous mappings is also uniformly continuous.

Lemma 3.3.1 *A uniformly continuous mapping* $f : (X, \mathcal{U}_X) \to (Y, \mathcal{U}_Y)$ *between quasi-uniform spaces is strongly extensional.*

Proof. If $f(x) \neq f(y)$, then without loss of generality we may assume that there exists $V \in \mathcal{U}_Y$ such that $(f(x), f(y)) \in \neg V$. By the uniform continuity of f, there exists $U \in \mathcal{U}_X$ such that $(f \times f)(U) \subset V$. Clearly, $(x, y) \in \neg U$, so $x \neq y$. ∎

Proposition 3.3.2 *A uniformly continuous mapping between quasi-uniform spaces is strongly continuous.*

Proof. Let f be a uniformly continuous mapping between the quasi-uniform spaces $(X, \mathcal{U}_X)$ and $(Y, \mathcal{U}_Y)$, and consider subsets S, T of X such that $f(S) \bowtie f(T)$. Pick $V \in \mathcal{U}_Y$ such that $f(S) \times f(T) \subset \sim V$. By the uniform continuity of f, there exists $U \in \mathcal{U}_X$ such that if $(x, y) \in U$, then $(f(x), f(y)) \in V$. Let $s \in S$, $t \in T$, and $(x, y) \in U$. Then $(f(s), f(t)) \in \sim V$ and $(f(x), f(y)) \in V$, so $(f(s), f(t)) \neq (f(x), f(y))$. Hence either $f(s) \neq f(x)$ or $f(t) \neq f(y)$. The strong extensionality of f (Lemma 3.3.1) now ensures that either $s \neq x$ or $t \neq y$, so $(s, t) \neq (x, y)$. Hence $S \times T \subset \sim U$, and so $S \bowtie T$. ∎

We shall discuss the converse of this proposition shortly; but first, in order to provide ourselves with a necessary and sufficient condition for uniform continuity, we introduce the notion of the product of two quasi-uniform structures. Given quasi-uniform spaces $(X_k, \mathcal{U}_k)$ and entourages $U_k \in \mathcal{U}_k$ ($k = 1, 2$), we define

$$U_1 \boxtimes U_2 \equiv \left\{ (\mathbf{x}, \mathbf{y}) \in (X_1 \times X_2)^2 : ((x_1, y_1), (x_2, y_2)) \in U_1 \times U_2 \right\}.$$

Sets of this type form a filter base on the Cartesian product $(X_1 \times X_2)^2$. We denote by $\mathcal{U}_1 \times \mathcal{U}_2$ the filter generated by sets of this form.

Proposition 3.3.3 *Let $(X_1, \mathcal{U}_1)$ and $(X_2, \mathcal{U}_2)$ be quasi-uniform (respectively, uniform) spaces. Then $\mathcal{U}_1 \times \mathcal{U}_2$ is a quasi-uniform (respectively, uniform) structure on the product space $X \equiv X_1 \times X_2$.*

Proof. Certainly, $\mathcal{U} \equiv \mathcal{U}_1 \times \mathcal{U}_2$ satisfies **U1**. Since the sets of the form $U_1 \boxtimes U_2$, with $U_1 \in \mathcal{U}_1$ and $U_2 \in \mathcal{U}_2$, generate the filter $\mathcal{U}_1 \times \mathcal{U}_2$, we see from **U2** in the factor spaces that

$$(\mathbf{x}, \mathbf{y}) \in \bigcap \{U : U \in \mathcal{U}_1 \times \mathcal{U}_2\} \Leftrightarrow \forall_{k \in \{1,2\}} \forall_{U_k \in X_k} ((x_k, y_k) \in U_k)$$

$$\Leftrightarrow \forall_{k \in \{1,2\}} (x_k = y_k)$$

$$\Leftrightarrow \mathbf{x} = \mathbf{y}.$$

This verifies **U2** in the product space,

Now consider any sets $U_k \in \mathcal{U}_k$. To verify **U3** in X, we use that axiom in X_k to pick $V_k \in \mathcal{U}_k$ such that $V_k^2 \subset U_k$. Let $(\mathbf{x}, \mathbf{y}) \in (V_1 \boxtimes V_2)^2$. Then there exists $\mathbf{z}$ such that $(\mathbf{x}, \mathbf{z}) \in V_1 \boxtimes V_2$ and $(\mathbf{z}, \mathbf{y}) \in V_1 \boxtimes V_2$. Hence

$$((x_1, z_1), (x_2, z_2)) \in V_1 \times V_2 \text{ and } ((z_1, y_1), (z_2, y_2)) \in V_1 \times V_2,$$

so $(x_k, z_k) \in V_k$ and $(z_k, y_k) \in V_k$. It follows that $(x_k, y_k) \in V_k^2 \subset U_k$; whence $(\mathbf{x}, \mathbf{y}) \in U_1 \boxtimes U_2$. Thus $(V_1 \boxtimes V_2)^2 \subset U_1 \boxtimes U_2$.

To verify **U4**, use that axiom in X_k to construct $V_k \in \mathcal{U}_k$ such that $X_k = U_k \cup \neg V_k$, and consider any $(\mathbf{x}, \mathbf{y}) \in X$. Either $(x_k, y_k) \in U_k$ for each k, or else there exists k such that $(x_k, y_k) \in \neg V_k$. In the first case we have $(\mathbf{x}, \mathbf{y}) \in U_1 \boxtimes U_2$; in the second,

$$((x_1, y_1), (x_2, y_2)) \in \neg (V_1 \times V_2)$$

and therefore $(\mathbf{x}, \mathbf{y}) \in \neg (V_1 \boxtimes V_2)$. This completes the proof that $\mathcal{U}$ is a quasi-uniform structure on X. It remains to observe that if each X_k is a uniform space, then, by Proposition 3.2.18, the sets of the type $U_1 \boxtimes U_2$ with U_k a symmetric entourage of X_k generate the filter $\mathcal{U}$; so $\mathcal{U}$ is a uniform structure. ∎

The (quasi-)uniform structure $\mathcal{U}_1 \times \mathcal{U}_2$ in Proposition 3.3.3 is called the **product of the (quasi-)uniform structures** on X_1 and X_2, or the **product (quasi-)uniform structure** on $X_1 \times X_2$. It is a simple consequence of the definition of this structure that the projection mappings $\mathrm{pr}_k : X_1 \times X_2 \to X_k$ are uniformly (and hence, by Proposition 3.3.2, strongly) continuous. Also, if X_1, X_2 are quasi-metric spaces, then the product quasi-uniform structure on X coincides with the quasi-uniform structure induced by the product quasi-metric on X.

Proposition 3.3.4 *A mapping f between uniform spaces $(X, \mathcal{U}_X)$ and $(Y, \mathcal{U}_Y)$ is uniformly continuous if and only if $f \times f$ is strongly continuous on the product uniform space $X \times X$.*

Proof. For convenience, write $F \equiv f \times f$. Supposing that f is uniformly continuous, for $k \in \{1,2\}$ let $V_k \in \mathcal{U}_Y$ and construct $U_k \in \mathcal{U}_X$ such that $(f \times f)(U_k) \subset V_k$. Then

$$(F \times F)(U_1 \boxtimes U_2) \subset V_1 \boxtimes V_2.$$

It follows from the definition of the product uniform structure that $f \times f$ is uniformly continuous, and therefore, by Proposition 3.3.2, strongly continuous on $X \times X$.

Now suppose, conversely, that F is strongly continuous, and let $V \in \mathcal{U}_Y$. Construct a 3-chain (V_1, V_2, V_3) of entourages of Y such that $V_1 = V$ and V_3 is symmetric. Setting $T \equiv F^{-1}(\neg V_2) \subset X \times X$, we show that $F(\Delta_X) \bowtie F(T)$ (where Δ_X is the diagonal of $X \times X$) in the product uniform space $Y \times Y$. Given $(x, x') \in T$ and $z \in X$, we see that if $(f(z), f(x)) \in V_3$ and $(f(z), f(x')) \in V_3$, then $(f(x), f(z)) \in V_3^{-1} = V_3$ and therefore $(f(x), f(x')) \in V_3 \circ V_3 \subset V_2$. This is impossible since $(x, x') \in T$. Hence

$$((f(z), f(x)), (f(z), f(x'))) \in \neg (V_3 \times V_3)$$

—that is,

$$(F(z, z), F(x, x')) \in \neg(V_3 \boxtimes V_3).$$

We conclude that

$$F(\Delta_X) \times F(T) \subset \neg(V_3 \boxtimes V_3)$$

and therefore that $F(\Delta_X) \bowtie F(T)$ in the product uniform space $Y \times Y$. Since F is strongly continuous, $\Delta_X \bowtie T$ in the product uniform space $X \times X$, so there exists $U \in \mathcal{U}_X$ with $\Delta_X \times T \subset \neg(U \boxtimes U)$. We complete the proof by showing that $(f \times f)(U) \subset V$. For any $(x, x') \in U$ we have either $(f(x), f(x')) \in V$ or else $(f(x), f(x')) \in \neg V_2$. In the latter case, $(x, x') \in T$, so

$$((x, x), (x, x')) \in \Delta_X \times T \subset \neg(U \boxtimes U).$$

This is impossible, since

$$(x, x, x, x') \in \Delta_X \boxtimes U \subset U \boxtimes U.$$

Hence $(f \times f)(x, x') \in V$. ∎

Our goal in the rest of this section is to discuss two partial converses of Proposition 3.3.2. We shall work from now on with uniform, rather than quasi-uniform, spaces.

Given a uniform space $(X, \mathcal{U})$ and $U \in \mathcal{U}$, by a U-**approximation** to X we mean an inhabited subset A of X such that

$$X = \bigcup_{x \in A} U[x].$$

We say that X is **totally bounded** if for each $U \in \mathcal{U}$, there exists a finitely enumerable U-approximation to X. For metric spaces, this notion of total boundedness is equivalent to the one introduced in Chapter 1.

By a U-**small set** in the uniform space X we mean a subset S of X such that $S \times S \subset U$. The property of total boundedness for X is equivalent to there being, for each $U \in \mathcal{U}$, a finitely enumerable family of inhabited U-small sets whose union is X.

By a **totally bounded subset** of a uniform space X we mean a subset that is totally bounded as a uniform subspace of X.

Proposition 3.3.5 *The closure (in the uniform topology) of a totally bounded subset of a uniform space is totally bounded.*

Proof. Let S be a totally bounded subset of the uniform space $(X, \mathcal{U})$. Given U in $\mathcal{U}$, pick a symmetric entourage V such that $V^2 \subset U$, and construct a finitely enumerable V-approximation $\{x_1, \ldots, x_n\}$ to S. Let $x \in \overline{S}$, and choose $y \in S$ such that $(x, y) \in V$. There exists i $(1 \leqslant i \leqslant n)$ such that $(x_i, y) \in V$. We then have $(x_i, x) \in V \circ V^{-1} = V^2 \subset U$. Hence $\{x_1, \ldots, x_n\}$ is a finitely enumerable U-approximation to $\overline{S}$. ∎

Proposition 3.3.6 *Let S be a totally bounded subset of $\mathbf{R}$. Then $\sup S$ and $\inf S$ exist.*

Proof. Let α, β be real numbers with $\alpha < \beta$, and set $\varepsilon \equiv \frac{1}{2}(\beta - \alpha)$. Pick a finitely enumerable subset $\{x_1, \ldots, x_n\}$ of S such that

$$S \subset \bigcup_{i=1}^{n} (x_i - \varepsilon, x_i + \varepsilon).$$

Either $x_i > \alpha$ for some i, or else $x_i < \alpha + \varepsilon$ for all i. In the latter case, given $x \in S$ and choosing i such that $|x - x_i| < \varepsilon$, we have

$$x \leqslant x_i + |x - x_i| < \alpha + \varepsilon + \varepsilon = \beta.$$

Thus either $x > \alpha$ for some $x \in S$, or else $x < \beta$ for all $x \in S$. The constructive least-upper-bound principle (see page 7) now shows that $\sup S$ exists. Hence

$$\inf S = -\sup\{-x : x \in S\}$$

also exists. ∎

Proposition 3.3.7 *Let X be a totally bounded uniform space, and f a uniformly continuous mapping of X into a uniform space Y. Then $f(X)$ is totally bounded.*

Proof. Let V be an entourage of Y, and pick an entourage U of X such that if $(x, x') \in U$, then $(f(x), f(x')) \in V$. There exist finitely many inhabited U-small sets $S_1, \ldots, S_n$ whose union is S. Then the inhabited sets $f(S_1), \ldots, f(S_n)$ are V-small, and their union is $f(X)$. ∎

Corollary 3.3.8 *Let f be a uniformly continuous mapping of a totally bounded uniform space X into $\mathbf{R}$. Then $\sup f$ and $\inf f$ exist.*

Proof. Apply Propositions 3.3.6 and 3.3.7. ∎

Proposition 3.3.9 *Let X and Y be uniform apartness spaces, with Y totally bounded, and let f be a strongly continuous mapping of X onto Y. Then f is uniformly continuous.*

Proof. Given an entourage V of Y, construct a 4-chain (V_1, V_2, V_3, V_4) of entourages of Y such that $V_2^3 \subset V_1 = V$, V_4 is symmetric, and $V_4^3 \subset V_3$. Choose $x_1, \ldots, x_m$ in X such that $Y = Y_1 \cup \cdots \cup Y_m$, where $Y_i \equiv V_4[f(x_i)]$; then set $X_i \equiv f^{-1}(Y_i)$. For $1 \leqslant i, j \leqslant m$ construct c_{ij} such that

$$\begin{aligned} c_{ij} = 0 &\Rightarrow (f(x_i), f(x_j)) \in V_2, \\ c_{ij} = 1 &\Rightarrow (f(x_i), f(x_j)) \in \neg V_3. \end{aligned}$$

We prove the following:

(i) if $c_{ij} = 0$, then $Y_i \times Y_j \subset V$;

(ii) if $c_{ij} = 1$, then $Y_i \bowtie Y_j$.

For (i), let $c_{ij} = 0$ and consider (x, x') with $(f(x), f(x')) \in Y_i \times Y_j$. We have $(f(x), f(x_i)) \in V_4^{-1} = V_4$ and $(f(x_j), f(x')) \in V_4$, so since $V_4 \subset V_2$,

$$(f(x), f(x')) \in V_4 \circ V_2 \circ V_4 \subset V_2^3 \subset V.$$

This proves (i). It follows that if $c_{ij} = 0$ for all i and j, then $(f(x), f(x')) \in V$ for all $x, x' \in X$. Thus we may assume that there exist i, j with $c_{ij} = 1$. For such i and j, consider an element (y, y') of $Y_i \times Y_j$, and suppose that $(y, y') \in V_4$. Then $(f(x_i), y) \in V_4$ and $(y', f(x_j)) \in V_4^{-1} = V_4$, so $(f(x_i), f(x_j)) \in V_4^3 \subset V_3$, a contradiction. Hence $(y, y') \in \neg V_4$. It follows that $Y_i \times Y_j \subset \neg V_4$ and therefore that $Y_i \bowtie Y_j$. This proves (ii). Moreover, we have $X_i \bowtie X_j$, by the strong continuity of f, so there exists an entourage U_{ij} of X with $X_i \times X_j \subset \neg U_{ij}$. Let

$$U \equiv \bigcap \{U_{ij} : c_{ij} = 1\},$$

which also is an entourage of X. Consider points x, x' of X with $(x, x') \in U$. Choose i, j such that $f(x) \in Y_i$ and $f(x') \in Y_j$. If $c_{ij} = 1$, then

$$(x, x') \in X_i \times X_j \subset \neg U_{ij} \subset \neg U,$$

a contradiction. Hence $c_{ij} = 0$ and therefore, by (i), $(f(x), f(x')) \in V$. Since $V \in \mathcal{U}_Y$ is arbitrary, this completes the proof that f is uniformly continuous. ∎

Corollary 3.3.10 *A given apartness space has at most one compatible uniformity that is totally bounded.*

Proof. Let $(X, \bowtie)$ be an apartness space, and suppose that there are two totally bounded uniformities $\mathcal{U}, \mathcal{U}'$ that are compatible with the apartness on X. Then the identity mapping from $(X, \bowtie)$ onto $(X, \bowtie_\mathcal{U})$ is strongly continuous, as is its inverse; likewise, the identity mapping from $(X, \bowtie)$ onto $(X, \bowtie_{\mathcal{U}'})$ is strongly continuous, as is its inverse. Hence the identity mapping from $(X, \bowtie_\mathcal{U})$ to $(X, \bowtie_{\mathcal{U}'})$ is strongly continuous, as is its inverse. It now follows from Proposition 3.3.9 that the identity mapping f from $(X, \mathcal{U})$ to $(X, \mathcal{U}')$ is uniformly continuous, as is its inverse. Thus for each $U' \in \mathcal{U}'$, there exists $U \in \mathcal{U}$ such that if $(x, y) \in U$, then $(x, y) = (f(x), f(y)) \in U'$; from which it follows that $U \subset U'$ and therefore, by **U1**, $U' \in \mathcal{U}$. Hence $\mathcal{U}' \subset \mathcal{U}$, and likewise $\mathcal{U} \subset \mathcal{U}'$. ∎

Classically, every symmetric (pre-)apartness space with the Efremovič property has exactly one totally bounded compatible uniformity. We have already seen, in Corollary 3.2.30, that the existence of *any* compatible uniformity for a certain set-set pre-apartness relation of that type implies the weak law of excluded middle.

Proposition 3.3.9 will be applied again later. In the meantime, we explore what can and cannot be proved by way of a converse to Proposition 3.3.2 when the range of the strongly continuous function is not totally bounded.

We say that two sequences $(x_n)_{n\geqslant 1}, (x'_n)_{n\geqslant 1}$ in a uniform space $(X, \mathcal{U})$ are **eventually close** if

$$\forall_{U \in \mathcal{U}} \exists_N \forall_{n \geqslant N} \left((x_n, x'_n) \in U \right).$$

A mapping f of X into a uniform space Y is **uniformly sequentially continuous** if the sequences $(f(x_n))_{n\geqslant 1}, (f(x'_n))_{n\geqslant 1}$ are eventually close in Y whenever $(x_n)_{n\geqslant 1}, (x'_n)_{n\geqslant 1}$ are eventually close in X. We shall prove the following partial converse of Proposition 3.3.2.

Theorem 3.3.11 *A strongly continuous mapping $f : X \to Y$ between uniform spaces is uniformly sequentially continuous.*

In order to do so, we introduce a new proof technique, based on the next two lemmas. For these, we recall the following definitions relating to a subset S of $\mathbf{N}$: we say that

- $n \in S$ **eventually**, or **for all sufficiently large** n, if there exists $N \in \mathbf{N}$ such that $n \in S$ whenever $n \geqslant N$;

- $n \in S$ **infinitely often** if for each $n \in \mathbf{N}$ there exists $m > n$ such that $m \in S$.

In the latter case, using dependent choice, we can construct a strictly increasing sequence $(n_k)_{k\geqslant 1}$ of positive integers such that $n_k \in S$ for each k.

Lemma 3.3.12 *Let S be an inhabited set, and H a set of sequences in S such that if $s \equiv (s_n)_{n\geqslant 1} \in H$, then each subsequence of s belongs to H. Let T be a subset of S with the following property: for all $P, Q \subset \mathbf{N}^+$, if*

$$s \in H, \ \mathbf{N}^+ = P \cup Q, \ \text{and} \ s_n \in T \ \text{for each} \ n \in Q, \qquad (3.20)$$

then either $n \in P$ for all n or else there exists $n \in Q$. If (3.20) obtains, then either $n \in P$ eventually or else $n \in Q$ infinitely often.

Proof. Suppose that (3.20) holds. Then either $n \in P$ for all n or else, as we may assume, there exists $n_1 \in Q$. Set $\lambda_1 \equiv 0$. Using dependent choice, we construct inductively an increasing binary sequence $(\lambda_k)_{k\geqslant 1}$ and a strictly increasing sequence $(n_k)_{k\geqslant 1}$ of positive integers such that for each $k \geqslant 1$,

▷ if $\lambda_k = 0$, then $n_k \in Q$, and

▷ if $\lambda_k = 1 - \lambda_{k-1}$, then $n \in P$ for all $n > n_{k-1}$.

To this end, let $k \geqslant 2$ and suppose we have found n_{k-1} with the applicable properties. We may assume that $\lambda_{k-1} = 0$. Define

$$P' \equiv \{j \geqslant 1 : j + n_{k-1} \in P\},$$
$$Q' \equiv \{j \geqslant 1 : j + n_{k-1} \in Q\},$$

and

$$s' \equiv \left(s_{j+n_{k-1}}\right)_{j \geqslant 1}.$$

Then $s' \in H$, by our hypothesis about H, and $\mathbf{N}^+ = P' \cup Q'$. Moreover, if $j \in Q'$, then $j + n_{k-1} \in Q$ and so $s'_j \in T$. Applying our hypotheses with P,Q,s replaced by P',Q',s' respectively, we see that either $n \in P$ for all $n > n_{k-1}$, in which case we set $n_{k-1+j} \equiv j + n_{k-1}$ and $\lambda_{k-1+j} \equiv 1$ for each $j \geqslant 1$; or else there exists $n_k > n_{k-1}$ with $n_k \in Q$, when we set $\lambda_k \equiv 0$. This completes our inductive construction.

Now let

$$P'' \equiv \left\{k \in \mathbf{N}^+ : \lambda_k = 0 \lor \lambda_{k-1} = 1\right\},$$
$$Q'' \equiv \left\{k \in \mathbf{N}^+ : \lambda_k = 1 - \lambda_{k-1}\right\},$$

and

$$s'' \equiv \left(s_{n_{k-1}}\right)_{k \geqslant 2}.$$

Then $s'' \in H$ and $\mathbf{N}^+ = P'' \cup Q''$. Moreover, if $k \in Q''$, then $n_{k-1} \in Q$ and $s''_k = s_{n_{k-1}} \in T$. Applying our hypotheses with P,Q,s_k replaced by P'',Q'',s''_k respectively, we see that either $k \in P''$ for all k or else there exists $k \in Q''$. In the first case, if $\lambda_j = 1 - \lambda_{j-1}$ for some j, then $j \notin P''$, a contradiction; whence $\lambda_k = 0$ for all k, and therefore $(n_k)_{k \geqslant 1}$ is a strictly increasing sequence of elements of Q. In the case where there exists $k \in Q''$, we have $n \in P$ for all $n > n_{k-1}$. ∎

Our next lemma may seem bizarre, since it shows that under certain hypotheses the nonconstructive proposition **LPO** holds. However, it enables us to use **LPO** to rule out the unwanted second alternative in the conclusion of Lemma 3.3.12. This idea is used on a number of occasions in constructive proofs; see, for example, pages 195–196 of [29].

Lemma 3.3.13 *Let S,H, and T be as in* Lemma 3.3.12. *Let $s \equiv (s_n)_{n \geqslant 1} \in H$, let $\mathbf{N}^+ = P \cup Q$, and suppose that $s_n \in T$ for each $n \in Q$. If $n \in Q$ infinitely often, then* **LPO** *holds.*

Proof. Choose a strictly increasing sequence $(n_k)_{k \geqslant 1}$ in Q. Given an increasing binary sequence $(\lambda_k)_{k \geqslant 1}$, define

$$P' \equiv \left\{k \in \mathbf{N}^+ : \lambda_k = 0\right\},$$
$$Q' \equiv \left\{k \in \mathbf{N}^+ : \lambda_k = 1\right\}.$$

Then $(s_{n_k})_{k \geqslant 1} \in H$, $\mathbf{N}^+ = P' \cup Q'$, and $s_{n_k} \in T$ for all $k \in Q'$ (in fact, for all k). It follows from the hypotheses that either $k \in P'$ for all k or else there exists $k \in Q'$; so either $\lambda_k = 0$ for all k or else there exists k with $\lambda_k = 1$. ∎

In order to apply the foregoing, we need to set up suitable S, H, T; show that if $s \in H$, $\mathbf{N}^+ = P \cup Q$, and $s_n \in T$ for each $n \in Q$, then either $n \in P$ for all n or else there exists $n \in Q$; and use **LPO** to show that it is impossible for Q to be infinite. It will then follow that $n \in P$ eventually. For the application to the proof of Theorem 3.3.11 we need a few more lemmas.

Lemma 3.3.14 *Let X, Y be uniform spaces, $f : X \to Y$ a strongly continuous function, and V an entourage of Y. Let $(\lambda_n)_{n \geqslant 1}$ be an increasing binary sequence, and $(A_n)_{n \geqslant 1}, (B_n)_{n \geqslant 1}$ sequences of subsets of X such that*

▷ *for each entourage U of X there exists N such that for each $n \geqslant N$, either $A_n \times B_n = \varnothing$ or else $A_n \times B_n$ intersects U;*

▷ *if $\lambda_n = 0$, then $A_n = \varnothing$; and*

▷ *if $\lambda_n = 1 - \lambda_{n-1}$, then A_n and B_n are inhabited, $f(A_n) \times f(B_n) \subset \neg V$, and $A_k = \varnothing$ for all $k > n$.*

Then there exists N such that $\lambda_n = \lambda_N$ for all $n \geqslant N$.

Proof. We may assume that $\lambda_1 = 0$. Writing

$$A \equiv \bigcup_{n \geqslant 1} A_n,$$
$$B \equiv \bigcup \{B_n : n > 1 \wedge \lambda_n = 1 - \lambda_{n-1}\},$$

we have

$$f(A) \times f(B) \subset \neg V. \tag{3.21}$$

For if $x \in A$, then there exists n such that $x \in A_n$; so $\lambda_n = 1 - \lambda_{n-1}$, from which we obtain $A = A_n \neq \varnothing$, $B = B_n \neq \varnothing$, and $f(A_n) \times f(B_n) \subset \neg V$. It follows from (3.21) that $f(A) \bowtie f(B)$ and therefore, by the strong continuity of f, that there exists an entourage U of X such that $A \times B \subset \neg U$. Choose N such that for each $n \geqslant N$, either $A_n \times B_n = \varnothing$ or else $A_n \times B_n$ intersects U. Either $\lambda_N = 1$ and therefore $\lambda_n = 1$ for all $n \geqslant N$, or else $\lambda_N = 0$. In the latter case, if $\lambda_m = 1 - \lambda_{m-1}$ for some $m > N$, then $A = A_m \neq \varnothing$, $B = B_m \neq \varnothing$, and $A \times B$ intersects U. This contradicts our choice of U; whence $\lambda_n = 0$ for all $n \geqslant N$. ∎

Recall here that, for sets A and B, B^A is the set of all mappings from A into B. In particular $B^{\mathbf{N}^+}$ is the set of all infinite sequences in B.

Lemma 3.3.15 *Let X, Y be uniform spaces, $f : X \to Y$ a strongly continuous function, and V an entourage of Y. Let S be the space $X \times X$,*

$$H \equiv \left\{ s \in S^{\mathbf{N}^+} : \forall_{U \in \mathcal{U}_X} \exists_N \forall_{n \geqslant N} (s_n \in U) \right\},$$

and

$$T \equiv \{ (x, x') \in S : (f(x), f(x')) \in \neg V \}.$$

If $s \in H$, $\mathbf{N}^+ = P \cup Q$, and $s_n \in T$ for each $n \in Q$, then either $n \in P$ for all n or else there exists $n \in Q$.

Proof. Suppose that $s \equiv (s_n)_{n \geqslant 1} \in H$, that $\mathbf{N}^+ = P \cup Q$, and that $s_n \equiv (x_n, x'_n) \in T$ for each $n \in Q$. We may assume that $1 \in P$. Setting $\lambda_1 \equiv 0$, construct an increasing binary sequence $(\lambda_n)_{n \geqslant 1}$ such that for $n \geqslant 1$,

- if $\lambda_n = 0$, then $k \in P$ for all $k \leqslant n$;

- if $\lambda_n = 1 - \lambda_{n-1}$, then $n \in Q$.

We construct sequences $(A_n)_{n \geqslant 1}, (B_n)_{n \geqslant 1}$ of subsets of X as follows. If $\lambda_n = 0$, set $A_n \equiv \varnothing$ and $B_n \equiv \varnothing$. If $\lambda_n = 1 - \lambda_{n-1}$, set $A_n \equiv \{x_n\}$ and $B_n \equiv \{x'_n\}$, and note that, as $n \in Q$, we have $(x_n, x'_n) \in T$ and therefore $f(A_n) \times f(B_n) \subset \neg V$; in this case, set $A_k \equiv \varnothing$ and $B_k \equiv \varnothing$ for each $k > n$. Now consider any entourage U of X. Since $s \in H$, there exists ν such that $(x_n, x'_n) \in U$ for all $n \geqslant \nu$. For each such n, if $\lambda_n = 0$ or $\lambda_{n-1} = 1$, then $A_n \times B_n = \varnothing$. On the other hand, if $\lambda_n = 1 - \lambda_{n-1}$, then $A_n \times B_n = \{(x_n, x'_n)\} \subset U$. Thus the hypotheses of Lemma 3.3.14 are satisfied. Applying that lemma, we produce N such that $\lambda_n = \lambda_N$ for all $n \geqslant N$. If $\lambda_N = 0$, then $n \in P$ for all n; whereas if $\lambda_N = 1$, there exists $k \leqslant N$ such that $k \in Q$. ∎

Lemma 3.3.16 *Let Y be a uniform space, let $(a_n)_{n \geqslant 1}, (b_n)_{n \geqslant 1}$ be sequences in Y, and let (V_1, V_2) be a 2-chain of entourages of Y such that V_2 is symmetric, $V_2^3 \subset V_1$, and $(a_n, b_n) \in \neg V_1$ for each n. Then it is impossible that for each n, $(a_n, b_k) \in V_2$ for all sufficiently large k.*

Proof. Suppose that for each n we have $(a_n, b_k) \in V_2$ for all sufficiently large k. Choose N such that $(a_1, b_k) \in V_2$ for all $k \geqslant N$. By our supposition, there exists M such that $(a_N, b_k) \in V_2$ for all $k \geqslant M$. Take $K \equiv \max\{M, N\}$. Then $(a_N, b_K) \in V_2$, $(a_1, b_K) \in V_2$, and $(a_1, b_N) \in V_2$; whence, as V_2 is symmetric, $(a_N, b_N) \in V_2^3 \subset V_1$, a contradiction. ∎

Lemma 3.3.17 *Assume **LPO**, and let (V_1, V_2, V_3) be a 3-chain of symmetric entourages of Y such that $V_2^3 \subset V_1$ and $(a_n, b_n) \in \neg V_1$ for each n. Then there exists a strictly increasing sequence $(n_k)_{k \geqslant 1}$ of positive integers such that $(a_{n_j}, b_{n_k}) \in \neg V_3$ for all j and k.*

Proof. By **LPO**, for each positive integer n, either $(a_n, b_j) \in V_2$ for all sufficiently large j or else $(a_n, b_j) \in \neg V_3$ for infinitely many j. Another application of **LPO** ensures that either there exists ν_1 such that $(a_{\nu_1}, b_j) \in \neg V_3$ for infinitely many j, or else for each n, $(a_n, b_j) \in V_2$ for all sufficiently large j. The second case is ruled out by Lemma 3.3.16. Picking ν_1 as in the first case, construct a strictly increasing sequence $(k_{1,j})_{j \geqslant 1}$ of positive integers such that $k_{1,1} > \nu_1$ and $(a_{\nu_1}, b_{k_{1,j}}) \in \neg V_3$ for each j. Using **LPO** twice, and then Lemma 3.3.16, as before, we can produce j' such that $\left(a_{k_{1,j'}}, b_{k_{1,j}}\right) \in \neg V_3$ for infinitely many j. Setting $\nu_2 \equiv k_{1,j'}$, we construct a subsequence $(k_{2,j})_{j \geqslant 1}$ of $(k_{1,j})_{j \geqslant 1}$ such that $k_{2,1} > \nu_2$ and $(a_{\nu_2}, b_{k_{2,j}}) \in \neg V_3$ for each j. Then $(a_{\nu_1}, b_{\nu_2}) \in \neg V_3$. Carrying on in this way, we produce a strictly increasing sequence $(\nu_i)_{i \geqslant 1}$ of positive integers such that $(a_{\nu_i}, b_{\nu_j}) \in \neg V_3$ whenever $i < j$. Since $\neg V_1 \subset \neg V_3$, this holds also for $i = j$.

Passing to a subsequence, we may assume that $(a_i, b_j) \in \neg V_3$ whenever $i \leqslant j$. Now apply the foregoing argument to the sequences $(b_k)_{k \geqslant 1}$ and $(a_k)_{k \geqslant 1}$, in that order. We obtain a strictly increasing sequence $(n_k)_{k \geqslant 1}$ of positive integers such that $(b_{n_k}, a_{n_j}) \in \neg V_3$, and therefore $(a_{n_j}, b_{n_k}) \in \neg V_3$, for all $k \leqslant j$. Since also $(a_{n_j}, b_{n_k}) \in \neg V_3$ for all $j \leqslant k$, we conclude that $(a_{n_j}, b_{n_k}) \in \neg V_3$ for all j and k. ∎

At long last we can prove Theorem 3.3.11.

Proof. Let $S \equiv X \times X$, and define H as in Lemma 3.3.15. Let V be any entourage of Y, and construct a 4-chain (V, V_1, V_2, V_3) of entourages of Y such that V_1, V_2, V_3 are symmetric and $V_2^3 \subset V_1$. Define

$$T \equiv \{(x, x') \in S : (f(x), f(x')) \in \neg V_1\}.$$

By Lemmas 3.3.15 and 3.3.12, for any subsets P, Q of $\mathbf{N}^+$, if $\mathbf{N}^+ = P \cup Q$ and $s_n \in T$ for all $n \in Q$, then either $n \in P$ eventually or else $n \in Q$ infinitely often. Given two sequences $(x_n)_{n \geqslant 1}, (x'_n)_{n \geqslant 1}$ in X that are eventually close, let $s \equiv ((x_n, x'_n))_{n \geqslant 1}$. Since for each $n \in \mathbf{N}^+$ either $(f(x_n), f(x'_n)) \in V$ or else $(f(x_n), f(x'_n)) \in \neg V_1$ and $s_n \in T$, we see that either $(f(x_n), f(x'_n)) \in V$ eventually or $(f(x_n), f(x'_n)) \in \neg V_1$ infinitely often. Assume that the latter alternative obtains. Then, by Lemma 3.3.13, we can derive **LPO**. It follows from Lemma 3.3.17 that there exists a strictly increasing sequence $(n_k)_{k \geqslant 1}$ of positive integers such that $(f(x_{n_j}), f(x'_{n_k})) \in \neg V_3$ for all j, k. Writing

$$A \equiv \{x_{n_j} : j \geqslant 1\}, \quad B \equiv \{x'_{n_k} : k \geqslant 1\},$$

we see that $f(A) \times f(B) \subset \neg V_3$, so $f(A) \bowtie f(B)$ in Y. Since f is strongly continuous, $A \bowtie B$ in X, and therefore there exists an entourage U of X such that $A \times B \subset \neg U$. But this is absurd, since $(x_{n_k}, x'_{n_k}) \in U$ eventually. We therefore conclude that $(f(x_n), f(x'_n)) \in V$ eventually. ∎

It is shown in [23] that Theorem 3.3.11 cannot be extended to produce the uniform continuity of f unless we employ a principle that, although valid in the intuitionistic, recursive, and classical models of constructive mathematics, is not derivable with only intuitionistic logic and dependent choice (see [66]). However, under stronger hypotheses on the domain, we can obtain the conclusion that a strongly continuous function between uniform spaces is not just uniformly sequentially continuous but uniformly continuous.

Our next aim is to prove a second converse of Proposition 3.3.2:

Theorem 3.3.18 *Let X be a totally bounded uniform space with a countable base of entourages. Then every strongly continuous mapping from X into a uniform space is uniformly continuous.*

Not surprisingly, this too will require a number of technical lemmas.

Lemma 3.3.19 *If a uniform space has a countable base of entourages, then it has a countable base $(U_n)_{n\geqslant 1}$ of symmetric entourages such that $U_{n+1} \subset U_n$ for each n.*

Proof. Let $(V_n)_{n\geqslant 1}$ be a countable base of entourages of the uniform space $(X,\mathcal{U})$. Define U_n recursively, as follows:

$$
U_n \equiv \begin{cases} V_1 \cap V_1^{-1} & \text{if } n = 1 \\ \\ U_{n-1} \cap V_n \cap V_n^{-1} & \text{if } n > 1. \end{cases}
$$

Then $U_n \in \mathcal{U}$, $U_n = U_n^{-1}$, and $U_{n+1} \subset U_n$ for each n. Given $U \in \mathcal{U}$, pick n such that $V_n \subset U$; then $V_n \cap V_n^{-1} \subset U$, from which it readily follows that $U_n \subset U$. ∎

Lemma 3.3.20 *A totally bounded uniform space $(X,\mathcal{U})$ with a countable base of entourages is separable. Moreover, if $(x_n)_{n\geqslant 1}$ is a dense sequence in X, then for each entourage U of X, there exists N such that*

$$ X = \bigcup_{i=1}^{N} U[x_i]. \tag{3.22} $$

Proof. Using Lemma 3.3.19, construct a countable base $(U_n)_{n\geqslant 1}$ of symmetric entourages for the totally bounded uniform space $(X,\mathcal{U})$. For each n, there exists an inhabited finitely enumerable subset F_n of X such that

$$ X = \bigcup_{y \in F_n} U_n[y]. $$

Let C be the countable set $\bigcup_{n\geqslant 1} F_n$. Given $x \in X$ and $U \in \mathcal{U}$, pick n such that $U_n \subset U$. There exists $y \in F_n \subset C$ such that $(x,y) \in U_n^{-1} = U_n \subset U$ and therefore $y \in C \cap U[x]$. It follows by the definition of the uniform topology on X that C is dense in X.

Now let $(x_n)_{n\geqslant 1}$ be a dense sequence in X, and U any entourage of X. Pick a symmetric entourage V such that (U,V) is a 2-chain. There exist finitely many points $\{y_1, \ldots, y_k\}$ of X such that $X = \bigcup_{i=1}^{k} V[y_i]$. For each i, there exists n_i such that $x_{n_i} \in V[y_i]$. Given $x \in X$, choose i such that $x \in V[y_i]$. Then $(x_{n_i}, x) \in V^{-1} \circ V \subset U$. Setting

$$N \equiv \max\{n_1, \ldots, n_k\},$$

we see that (3.22) holds. ∎

Lemma 3.3.21 *Let X be a totally bounded uniform space with a countable base of entourages, and f a strongly continuous mapping of X into a uniform space Y. Let $(x_n)_{n \geqslant 1}$ be a dense sequence in X (which exists by Lemma 3.3.20), and V any entourage of Y. Let $(\lambda_n)_{n \geqslant 1}$ be an increasing binary sequence, $(k_n)_{n \geqslant 1}$ a strictly increasing sequence of positive integers, and $(A_n)_{n \geqslant 1}, (B_n)_{n \geqslant 1}$ sequences of subsets of X such that*

▷ *if $\lambda_n = 0$, then $A_n = \varnothing$, and*

▷ *if $\lambda_n = 1 - \lambda_{n-1}$, then A_n is inhabited, $B_n = \{x_1, \ldots, x_{k_{n-1}}\}$, $f(A_n) \times f(B_n) \subset \neg V$, and $A_k = \varnothing$ for all $k > n$.*

Then there exists N such that $\lambda_n = \lambda_N$ for all $n \geqslant N$.

Proof. Given an entourage U of X, use Lemma 3.3.20 to compute $N > 1$ such that

$$X = \bigcup_{i=1}^{k_{N-1}} U^{-1}[x_i],$$

and consider any $n \geqslant N$. If either $\lambda_n = 0$ or $\lambda_{n-1} = 1$, then $A_n \times B_n = \varnothing$. If $\lambda_n = 1 - \lambda_{n-1}$, then A_n is inhabited, $B_n = \{x_1, \ldots, x_{k_{n-1}}\}$, and for each $x \in A_n$ there exists i such that $1 \leqslant i \leqslant k_{N-1} \leqslant k_{n-1}$ and $x \in U^{-1}[x_i]$; so $(x, x_i) \in (A_n \times B_n) \cap U$. We can now apply Lemma 3.3.14 in order to derive the desired conclusion. ∎

Lemma 3.3.22 *If $f : X \to Y$ is a strongly continuous mapping between uniform spaces, then $f(\overline{S}) \subset \overline{f(S)}$ for each $S \subset X$.*

Proof. Corollary 3.2.27 shows that Y has the weak nested neighbourhoods property. The desired conclusion now follows from Corollary 2.3.8, Proposition 2.3.4, and Proposition 2.3.3. ∎

Lemma 3.3.23 *Let X be a totally bounded uniform space with a countable base of entourages. Let f be a strongly continuous mapping of X into a uniform space Y, let $(V_1, \ldots, V_5)$ be a 5-chain of entourages of Y, and let S be a finitely enumerable subset of X. Then*

▷ *either for each $x \in X$ there exists $s \in S$ such that $(f(x), f(s)) \in V_1$*

▷ *or else there exists $x \in X$ such that $\{f(x)\} \times f(S) \subset \neg V_5$.*

Proof. By Lemma 3.3.20, X has a dense sequence $(x_n)_{n\geqslant 1}$. Setting $\lambda_0 \equiv 0$, construct an increasing binary sequence $(\lambda_n)_{n\geqslant 1}$ such that for each $n \geqslant 1$,

$$\lambda_n = 0 \Rightarrow \forall_{k\leqslant n}\exists_{s\in S}\left((f(x_k),f(s))\in V_4\right),$$
$$\lambda_n = 1 - \lambda_{n-1} \Rightarrow \forall_{s\in S}\left((f(x_n),f(s))\in \neg V_5\right).$$

Clearly, we may assume that $\lambda_1 = 0$. If $\lambda_n = 0$, set $B_n \equiv \varnothing$; if $\lambda_n = 1-\lambda_{n-1}$, set $B_n \equiv \{x_1,\ldots,x_{n-1}\}$ and $B_k \equiv \varnothing$ for all $k > n$. Let $B \equiv \bigcup_{n\geqslant 1} B_n$.

Next, setting $\mu_0 \equiv 0$, construct an increasing binary sequence $(\mu_n)_{n\geqslant 1}$ such that for each $n \geqslant 1$,

$$\mu_n = 0 \Rightarrow \forall_{k\leqslant n}\exists_{s\in S}\left((f(x_k),f(s))\in V_2\right),$$
$$\mu_n = 1 - \mu_{n-1} \Rightarrow \forall_{s\in S}\left((f(x_n),f(s))\in \neg V_3\right).$$

If $\mu_n = 0$, set $A_n \equiv \varnothing$; if $\mu_n = 1 - \mu_{n-1}$, set $A_n \equiv \{x_n\}$ and $A_k \equiv \varnothing$ for all $k > n$. Let $A \equiv \bigcup_{n\geqslant 1} A_n$. We show that

$$f(A) \times f(B) \subset \neg V_4. \tag{3.23}$$

Since $\neg V_3 \subset \neg V_5$, we may assume that $\mu_1 = 0$. Let $x \in A$ and $y \in B$, and suppose that $(f(x),f(y)) \in V_4$. Choosing n such that $y \in B_n$, we see that $\lambda_n = 1 - \lambda_{n-1}$, $B = B_n = \{x_1,\ldots,x_{n-1}\}$, $y = x_j$ for some $j \leqslant n-1$, and there exists $s \in S$ such that $(f(y),f(s)) \in V_4$. Then

$$(f(x),f(s)) \in V_4^2 \subset V_3. \tag{3.24}$$

On the other hand, since $x \in A$, there exists k such that $\mu_k = 1 - \mu_{k-1}$, $x = x_k$, and $(f(x),f(s)) \in \neg V_3$, which contradicts (3.24). Thus $(f(x),f(y)) \in \neg V_4$. Since x,y are arbitrary elements of A,B respectively, we now obtain (3.23). Hence $f(A) \bowtie f(B)$.

It follows from the strong continuity of f that there exists a symmetric entourage U of X such that $A \times B \subset \neg U$. Since X is totally bounded, by Lemma 3.3.20 there exists N such that $X = \bigcup_{i=1}^{N} U[x_i]$. If $\lambda_N = 1$, then there exists $j \leqslant N$ such that $\lambda_j = 1 - \lambda_{j-1}$ and therefore $\{f(x_j)\} \times f(S) \subset \neg V_5$. So without loss of generality we may assume that $\lambda_N = 0$. Suppose that $\mu_n = 1 - \mu_{n-1}$ for some n. Then for each $s \in S$ we have $(f(x_n),f(s)) \in \neg V_3$ and therefore $(f(x_n),f(s)) \notin V_4$. Hence $\lambda_n \neq 0$, and therefore $\lambda_m = 1 - \lambda_{m-1}$ for some $m \leqslant n$; clearly, $N < m$. We now have $A = \{x_n\}$, $B = \{x_1,\ldots,x_{m-1}\}$, and

$$\{x_n\} \times \{x_1,\ldots,x_{m-1}\} \subset \neg U.$$

This is absurd since, by our choice of N, there exists $i \leqslant N < m$ such that $(x_i,x_n) \in U$ and therefore $(x_n,x_i) \in U$. It follows that for all n, $\mu_n = 0$ and therefore there exists $s \in S$ such that $(f(x_n),f(s)) \in V_2$. Now consider any $x \in X$. Since x is in the closure of $\{x_n : n \geqslant 1\}$, we see from Lemma 3.3.22

that $f(x)$ is in the closure of $\{f(x_n) : n \geqslant 1\}$; whence there exists n such that $f(x_n) \in V_2[f(x)]$. Choosing $s \in S$ such that $(f(x_n), f(s)) \in V_2$, we have $(f(x), f(s)) \in V_2^2 \subset V_1$. ∎

This brings us to the proof of Theorem 3.3.18.

Proof. Let X be a totally bounded uniform space with a countable base $(U_n)_{n \geqslant 1}$ of entourages. We may assume that each U_n is symmetric and that $U_1 \supset U_2 \supset \cdots$. By Lemma 3.3.20, there exists a dense sequence $(x_k)_{k \geqslant 1}$ in X; according to the same lemma, there exists a strictly increasing sequence $(k_n)_{n \geqslant 1}$ of positive integers such that

$$X = \bigcup_{j=1}^{k_n} U_n[x_j]$$

for each n. Let $F_n \equiv \{x_1, \ldots, x_{k_n}\}$. Consider a strongly continuous mapping f of X into a uniform space Y. In view of Proposition 3.3.9, it is enough to prove that $f(X)$ is totally bounded. Accordingly, given an entourage V of Y, construct a 5-chain $(V_1, \ldots, V_5)$ of entourages of Y with $V_1 = V$. Setting $\lambda_0 \equiv 0$ and using Lemma 3.3.23, construct an increasing binary sequence $(\lambda_n)_{n \geqslant 1}$ such that for each $n \geqslant 1$,

$$\lambda_n = 0 \Rightarrow \exists_{x \in X} \forall_{k \leqslant k_n} ((f(x), f(x_k)) \in \neg V_5),$$
$$\lambda_n = 1 - \lambda_{n-1} \Rightarrow \forall_{x \in X} \exists_{k \leqslant k_n} ((f(x), f(x_k)) \in V_1).$$

Since V is an arbitrary entourage of Y, it suffices to prove that there exists N with $\lambda_N = 1$: for in that case,

$$f(X) \subset \bigcup_{j=1}^{k_N} V[f(x_j)].$$

We may assume that $\lambda_1 = 0$. If $\lambda_n = 0$, set $A_n = B_n \equiv \varnothing$. If $\lambda_n = 1 - \lambda_{n-1}$, then, since $\lambda_{n-1} = 0$, there exists $\zeta \in X$ such that $(f(\zeta), f(x_k)) \in \neg V_5$ for all $k \leqslant k_{n-1}$; set $A_n \equiv \{\zeta\}$, $B_n \equiv F_{n-1}$, and $A_j = B_j \equiv \varnothing$ for all $j > n$. We have $f(A_n) \times f(B_n) \subset \neg V_5$ for all n. Now applying Lemma 3.3.21, compute N such that $\lambda_n = \lambda_N$ for all $n \geqslant N$. Assume that $\lambda_N = 0$. Then for each n, $\lambda_n = 0$ and so there exists $z_n \in X$ such that

$$\{f(z_n)\} \times f(F_n) \subset \neg V_5. \tag{3.25}$$

On the other hand, for each n there exists $\zeta_n \in F_n$ such that $(\zeta_n, z_n) \in U_n$ and therefore $(z_n, \zeta_n) \in U_n$. It follows that the sequences $(z_n)_{n \geqslant 1}$ and $(\zeta_n)_{n \geqslant 1}$ are eventually close: for if U is any entourage of X and we choose m such that $U_m \subset U$, then we have $(z_n, \zeta_n) \in U_n \subset U_m \subset U$ for all $n \geqslant m$. By Theorem 3.3.11, the sequences $(f(z_n))_{n \geqslant 1}$ and $(f(\zeta_n))_{n \geqslant 1}$ are eventually close, so there exists n such that $(f(z_n), f(\zeta_n)) \in V_5$; since $\zeta_n \in F_n$, this contradicts (3.25). Hence $\lambda_N \neq 0$, so $\lambda_N = 1$ and the proof is complete. ∎

We end this section with a diagram summarising the connections between strong and uniform continuity for mappings between uniform spaces.

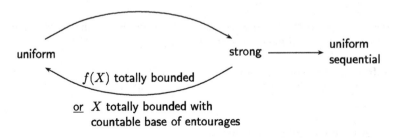

uniform ⟷ strong ⟶ uniform sequential

$f(X)$ totally bounded

<u>or</u> X totally bounded with countable base of entourages

3.4   Convergence of Nets of Functions

Let X be an inhabited set, and $(Y, \bowtie)$ a pre-apartness space. We say that a net $(f_n)_{n \in \mathfrak{D}}$ in Y^X is **proximally convergent** to $f \in Y^X$ if for all $A \subset X$ and $B \subset Y$ we have both

$$f(A) \bowtie B \Rightarrow \exists_N \forall_{n \geqslant N} (f_n(A) \bowtie B)$$

and

$$B \bowtie f(A) \Rightarrow \exists_N \forall_{n \geqslant N} (B \bowtie f_n(A)).$$

If the apartness on Y is symmetric, then we need bother only with the first of these two requirements. Proximal convergence of the net $(f_n)_{n \in \mathfrak{D}}$ to f implies **pointwise convergence**: for each x in X, the net $(f_n(x))_{n \in \mathfrak{D}}$ converges to $f(x)$ in Y.

When is the limit of a proximally convergent net of strongly continuous mappings also strongly continuous?

Proposition 3.4.1 *Let X be a pre-apartness space, Y a pre-apartness space with the Efremovič property, and $(f_n)_{n \in \mathfrak{D}}$ a net of strongly continuous mappings of X into Y that converges proximally to $f \in Y^X$. Then f is strongly continuous.*

Proof. Let A, B be subsets of X with $f(A) \bowtie f(B)$. Construct $E \subset Y$ such that $f(A) \bowtie \neg E$ and $E \bowtie f(B)$. Then there exist $n_1, n_2 \in \mathfrak{D}$ such that $f_n(A) \bowtie \neg E$ for all $n \succcurlyeq n_1$, and $E \bowtie f_n(B)$ for all $n \succcurlyeq n_2$. Pick $N \in \mathfrak{D}$ with $N \succcurlyeq n_1$ and $N \succcurlyeq n_2$. Then (see Lemma 3.1.3) $f_N(B) \subset {\sim}E \subset \neg E$; whence $f_N(A) \bowtie f_N(B)$. It follows from the strong continuity of f_N that $A \bowtie B$. ∎

Corollary 3.4.2 *Let X be a compact metric space, Y a uniform apartness space, and $(f_n)_{n \in \mathfrak{D}}$ a net of strongly continuous mappings from X into Y that converges proximally to $f \in Y^X$. Then f is uniformly continuous.*

Proof. By Proposition 3.4.1, f is strongly continuous; whence, by Theorem 3.3.18, it is uniformly continuous. ■

When the pre-apartness on Y is induced by a uniform structure $\mathcal{U}$, we have two other notions of convergence of a net $(f_n)_{n \in \mathfrak{D}}$ to f in Y^X:

▷ **uniform sequential convergence**, in which $\mathfrak{D} = \mathbf{N}^+$ and for each sequence $(x_n)_{n \geqslant 1}$ in X, the sequences $(f_n(x_n))_{n \geqslant 1}$ and $(f(x_n))_{n \geqslant 1}$ are eventually close;

▷ **uniform convergence**, in which

$$\forall_{U \in \mathcal{U}} \exists_N \forall_{n \succcurlyeq N} \forall_{x \in X} \left((f_n(x), f(x)) \in U \right).$$

It is almost trivial that uniform convergence implies uniform sequential convergence.

Proposition 3.4.3 *Let X be an inhabited set, Y a uniform space, and $(f_n)_{n \in \mathfrak{D}}$ a net in Y^X that converges uniformly to $f \in Y^X$. Then $(f_n)_{n \in \mathfrak{D}}$ converges proximally to f.*

Proof. Let $A \subset X$, $B \subset Y$, and $f(A) \bowtie B$. There exists a 2-chain (U, V) of entourages of Y such that V is symmetric and $f(A) \times B \subset \neg U$. Pick $N \in \mathfrak{D}$ such that $(f_n(x), f(x)) \in V$ for all $x \in X$ and all $n \succcurlyeq N$. Given such a value of n, x in A, and y in B, we see that if $(f_n(x), y) \in V$, then

$$(f(x), y) \in V^{-1} \circ V = V^2 \subset U,$$

a contradiction. Hence $f_n(A) \times B \subset \neg V$, and therefore $f_n(A) \bowtie B$, for all $n \succcurlyeq N$. ■

We shall prove later, in Corollary 3.4.16, that if Y is a totally bounded uniform space, then for nets of functions from X to Y, proximal convergence implies, and hence is equivalent to, uniform convergence. In the meantime, though, we aim to connect proximal and uniform sequential convergence for sequences.

Theorem 3.4.4 *Let X be an inhabited set, Y a uniform space, and $(f_n)_{n \geqslant 1}$ a sequence in Y^X that converges proximally to $f \in Y^X$. Then $(f_n)_{n \geqslant 1}$ is uniformly sequentially convergent to f.*

To do this, we need counterparts of Lemmas 3.3.14 and 3.3.15.

Lemma 3.4.5 *Let X be an inhabited set, Y a uniform space, and V an entourage of Y. Let $(f_n)_{n \geqslant 1}$ be a sequence in Y^X that converges proximally to $f \in Y^X$. Let $(\lambda_n)_{n \geqslant 1}$ be an increasing binary sequence and $(A_n)_{n \geqslant 1}$ a sequence of subsets of X such that*

> *if $\lambda_n = 0$ or $\lambda_{n-1} = 1$, then $A_n = \varnothing$, and*

> *if $\lambda_n = 1 - \lambda_{n-1}$, then there exists $x \in X$ such that $A_n = \{x\}$ and $(f(x), f_n(x)) \in \neg V$.*

Then there exists N such that $\lambda_n = \lambda_N$ for all $n \geqslant N$.

Proof. Write

$$A \equiv \bigcup_{n \geqslant 1} A_n,$$
$$B \equiv \bigcup \{ f_n(A_n) : n \geqslant 1 \wedge \lambda_n = 1 - \lambda_{n-1} \}.$$

If $x \in A$, then there exists n such that $A = A_n = \{x\}$, $B = \{f_n(x)\}$, and

$$f(A) \times B = \{(f(x), f_n(x))\} \subset \neg V.$$

It follows that $f(A) \times B \subset \neg V$ and therefore that $f(A) \bowtie B$. By proximal convergence, there exists N such that $f_n(A) \bowtie B$ for all $n \geqslant N$. Suppose that $\lambda_m = 1 - \lambda_{m-1}$ for some $m > N$. Then there exists $x \in X$ such that $A = A_m = \{x\}$ and $B = B_m = \{f_m(x)\}$; since $m > N$, we have $f_m(A) \bowtie B$ and therefore $\{f_m(x)\} \bowtie \{f_m(x)\}$, which is absurd. Hence $\lambda_n = \lambda_N$ for all $n \geqslant N$. ∎

Lemma 3.4.6 *Let X be an inhabited set, and $(Y, \mathcal{U})$ a uniform space. Let $f \in Y^X$, $S \equiv X \times Y^X$, and*

$$H \equiv \left\{ ((x_n, g_n))_{n \geqslant 1} \in S^{\mathbb{N}^+} : (g_n)_{n \geqslant 1} \text{ converges proximally to } f \right\}.$$

Let V be an entourage of Y, and

$$T \equiv \{ (x, g) \in S : (f(x), g(x)) \in \neg V \}.$$

If $s \equiv (s_n)_{n \geqslant 1} \in H$, $\mathbb{N}^+ = P \cup Q$, and $s_n \in T$ for each $n \in Q$, then either $n \in P$ eventually or else $n \in Q$ infinitely often.

Proof. Suppose that $s \in H$, $\mathbb{N}^+ = P \cup Q$, and $s_n \equiv (x_n, f_n) \in T$ for each $n \in Q$. In view of Lemma 3.3.12, it will suffice to prove that either $n \in P$ for all n or else there exists $n \in Q$. Setting $\lambda_0 \equiv 0$, construct an increasing binary sequence $(\lambda_n)_{n \geqslant 1}$ such that for each $n \geqslant 1$,

$$\lambda_n = 0 \Rightarrow \forall_{k \leqslant n} (k \in P),$$
$$\lambda_n = 1 - \lambda_{n-1} \Rightarrow n \in Q.$$

We may assume that $\lambda_1 = 0$. Define a sequence $(A_n)_{n \geqslant 1}$ of subsets of X as follows. If $\lambda_n = 0$ or $\lambda_{n-1} = 1$, set $A_n \equiv \varnothing$; if $\lambda_n = 1 - \lambda_{n-1}$, set $A_n \equiv \{x_n\}$ and note that, since $n \in Q$ and therefore $s_n \in T$, we have $(f(x_n), f_n(x_n)) \in \neg V$. By Lemma 3.4.5, there exists N such that $\lambda_n = \lambda_N$ for all $n \geqslant N$. If $\lambda_N = 0$, then $n \in P$ for all n; if $\lambda_N = 1$, then $n \in Q$ for some $n \leqslant N$. ∎

We now prove Theorem 3.4.4.

Proof. Under the hypotheses of Theorem 3.4.4, let $(x_n)_{n \geqslant 1}$ be any sequence in X. Let V be an entourage of Y, and construct a 4-chain (V, V_1, V_2, V_3) of entourages of Y with V_1, V_2, V_3 symmetric and $V_2^3 \subset V_1$. Let $S, H,$ and T be as in Lemma 3.4.6. Applying Lemma 3.4.6, we see that for all $P, Q \subset \mathbf{N}^+$, if $s \in H$, $\mathbf{N}^+ = P \cup Q$, and $s_n \in T$ for all $n \in Q$, then either $n \in P$ eventually or else $n \in Q$ infinitely often. Taking

$$P \equiv \{n : (f(x_n), f_n(x_n)) \in V\},$$
$$Q \equiv \{n : (f(x_n), f_n(x_n)) \in \neg V_1\},$$
$$s \equiv ((x_n, f_n))_{n \geqslant 1},$$

we see that either $(f(x_n), f_n(x_n)) \in V$ eventually or there exists a strictly increasing sequence $(n_k)_{k \geqslant 1}$ of positive integers such that $(f(x_{n_k}), f_{n_k}(x_{n_k})) \in \neg V_1$ for all k. Assume the latter alternative. Then **LPO** holds, by Lemma 3.3.13; so we can apply Lemma 3.3.17 with $a_k \equiv f(x_{n_k})$ and $b_k \equiv f_{n_k}(x_{n_k})$, to construct a strictly increasing sequence $(k_i)_{i \geqslant 1}$ of positive integers such that

$$\left(f(x_{n_{k_i}}), f_{n_{k_j}}(x_{n_{k_j}}) \right) \in \neg V_3 \quad (i, j \geqslant 1). \tag{3.26}$$

Writing

$$A \equiv \left\{ x_{n_{k_i}} : i \geqslant 1 \right\},$$
$$B \equiv \left\{ f_{n_{k_j}}(x_{n_{k_j}}) : j \geqslant 1 \right\},$$

we see from (3.26) and Proposition 3.2.11 that $f(A) \bowtie B$. Hence there exists N such that $f_n(A) \bowtie B$ for all $n \geqslant N$. In particular, using **B2**, we obtain $f_{n_{k_N}}(x_{n_{k_N}}) \neq f_{n_{k_N}}(x_{n_{k_N}})$. We conclude from this contradiction that $(f(x_n), f_n(x_n)) \in V$ eventually. Since V is an arbitrary entourage of Y, it follows that $(f_n)_{n \geqslant 1}$ is uniformly sequentially convergent to f. ∎

Are there apartness structures or topologies on Y^X relative to which the convergence of nets is none other than proximal convergence or uniform convergence? In answering this question, we deal only with the case where the pre-apartness on Y is symmetric.

For all $A \subset X$ and $B \subset Y$ we define

$$U_{A,B} \equiv \{f \in Y^X : f(A) \bowtie B\} \, .$$

The sets $U_{A,B}$, with $A \subset X$ and $B \subset Y$, form a subbase of a topology τ_p, called the **topology of proximal convergence**, on Y^X. We denote by $\bowtie_{Y^X}$ the corresponding point-set pre-apartness, and by $-_{Y^X}$ the associated apartness complement; thus if $f \in Y^X$ and $S \subset Y^X$, then $f \bowtie_{Y^X} S$ means that there exist finitely many subsets $A_1, \ldots, A_m$ of X, and finitely many subsets $B_1, \ldots, B_m$ of Y, such that

$$f \in \bigcap_{i=1}^{m} U_{A_i, B_i} \subset \sim S.$$

The following Brouwerian example shows that we cannot prove constructively that **A5** holds for the pre-apartness $\bowtie_{Y^X}$, even when Y is a two-point apartness space. Let $X \equiv \mathbf{N}$ and $Y \equiv \{0, 1\}$ (with the standard apartness), and let $f \in Y^X$ be the constant function 1. Given an increasing binary sequence $(a_n)_{n \geqslant 0}$, define

$$g(n) \equiv \min \{f(n), 1 - a_n\} \, .$$

Define also

$$S \equiv \{h \in Y^X : \exists_N \forall_{n \geqslant N} (h(n) = 0)\} \, .$$

Then $f \in U_{X, \{0\}} \subset \sim S$, so $f \bowtie_{Y^X} S$. Assuming **A5** for $\{0, 1\}^{\mathbf{N}}$, we see that either $f \neq g$ and therefore there exists n with $a_n = 1$, or else $g \bowtie_{Y^X} S$. In the latter case, $a_n = 0$ for all n: for if $a_n = 1 - a_{n-1}$, then $g \in S$. Thus if $\{0, 1\}^{\mathbf{N}}$ satisfies **A5**, then we can derive **LPO**.

Note that if a net $(f_n)_{n \in \mathfrak{D}}$ in Y^X converges to f relative to the topology τ_p, then it converges to f relative to the pre-apartness $\bowtie_{Y^X}$.

For the record, we prove a couple of simple results about separation properties of the pre-apartness on Y^X.

Proposition 3.4.7 Let X be an inhabited set, and Y a symmetric $\mathbf{T}_1$ pre-apartness space. Then $\left(Y^X, \bowtie_{Y^X}\right)$ is $\mathbf{T}_1$.

Proof. If $f, g \in Y^X$ and $f \neq g$, then there exists $x \in X$ such that $f(x) \neq g(x)$; so, since the apartness space Y is $\mathbf{T}_1$, $f(x) \bowtie g(x)$ and therefore $f \in U_{\{x\}, \{g(x)\}}$. But if $h \in U_{\{x\}, \{g(x)\}}$, then $h(x) \bowtie g(x)$ and therefore, by **A2** in Y, $h(x) \neq g(x)$. Hence $U_{\{x\}, \{g(x)\}} \subset \sim \{g\}$ and therefore $f \bowtie_{Y^X} g$. ∎

Proposition 3.4.8 Let X be an inhabited set, and Y a symmetric Hausdorff pre-apartness space. Then $\left(Y^X, \bowtie_{Y^X}\right)$ is Hausdorff.

Proof. If $f \neq g$ in Y^X, then we can find $x \in X$ such that $f(x) \neq g(x)$. Since Y is Hausdorff, there exist $S, T \subset Y$ such that $f(x) \in -S, g(x) \in -T$, and $-S \subset \sim -T$. Then

$$f \in U_{\{x\}, S} \subset \sim \sim U_{\{x\}, S}$$

and therefore $f \in -\sim U_{\{x\},S}$. Similarly, $g \in -\sim U_{\{x\},T}$. Moreover, for each $h \in U_{\{x\},S}$ we have

$$h(x) \in -S \subset \sim -T;$$

so $h(x) \neq h'(x)$ for each $h' \in U_{\{x\},T}$, and therefore $h \in \sim U_{\{x\},T}$. Thus $U_{\{x\},S} \subset \sim U_{\{x\},T}$. It follows that

$$-\sim U_{\{x\},T} \subset -U_{\{x\},S} \subset \sim U_{\{x\},S}$$

and hence that

$$-\sim U_{\{x\},S} \subset -\left(-\sim U_{\{x\},T}\right) \subset \sim\left(-\sim U_{\{x\},T}\right).$$

Thus Y^X is Hausdorff. ∎

Next, we explore the connection between proximal convergence on the one hand, and convergence relative to the topology τ_p and the pre-apartness $\bowtie_{Y^X}$ on the other.

Proposition 3.4.9 *Let X be an inhabited set, and Y a symmetric pre-apartness space. Let $(f_n)_{n\in\mathfrak{D}}$ be a net in Y^X, and $f \in Y^X$. Then $(f_n)_{n\in\mathfrak{D}}$ converges to f proximally if and only if it converges to f in the topology τ_p.*

Proof. Suppose that $(f_n)_{n\in\mathfrak{D}}$ converges proximally to f in Y^X, and let E be a τ_p-open set containing f. Then there exist subsets $A_1,\ldots,A_m$ of X, and subsets $B_1,\ldots,B_m$ of Y, such that

$$f \in \bigcap_{i=1}^{m} U_{A_i,B_i} \subset E.$$

For each i $(1 \leqslant i \leqslant m)$ there exists $n_i \in \mathfrak{D}$ such that $f_n \in U_{A_i,B_i}$ whenever $n \succcurlyeq n_i$. Choosing $N \in \mathfrak{D}$ such that $N \succcurlyeq n_i$ for each i, we see that if $n \succcurlyeq N$, then

$$f_n \in \bigcap_{i=1}^{m} U_{A_i,B_i} \subset E.$$

Thus $(f_n)_{n\in\mathfrak{D}}$ is τ_p-convergent to f.

Conversely, if $(f_n)_{n\in\mathfrak{D}}$ converges to f relative to the topology τ_p, and if $f(A) \bowtie B$, then f belongs to the τ_p-open set $U_{A,B}$; so there exists $N \in \mathfrak{D}$ such that $f_n \in U_{A,B}$—that is, $f_n(A) \bowtie B$—for all $n \succcurlyeq N$. ∎

Corollary 3.4.10 *Under the hypotheses of Proposition 3.4.9, proximal convergence in Y^X implies $\bowtie_{Y^X}$-convergence; but the converse holds if and only if (Y^X, τ_p) is topologically consistent.*

Proof. By Proposition 3.4.9, proximal convergence is equivalent to convergence in the topology τ_p, which, as we noted on page 111, implies convergence with respect to the corresponding pre-apartness $\bowtie_{Y^X}$. It remains to invoke Proposition 2.4.3. ∎

In order to discuss a topology associated with uniform convergence, we introduce a weakening of the notion of *uniform space*. Let E be an inhabited set. A filter $\mathcal{W}$ of subsets of $E \times E$ is called a **pre-uniform structure** on E if

- it satisfies **U1–U3** and

- $W^{-1} \in \mathcal{W}$ whenever $W \in \mathcal{W}$.

We then define the corresponding inequality on E as for a quasi-uniform space:

$$x \neq y \Leftrightarrow \exists_{W \in \mathcal{W}} ((x,y) \in \neg W).$$

Returning to the situation where X is an inhabited set and Y is a pre-apartness space, we now assume that the pre-apartness on Y is induced by a uniform structure $\mathcal{U}$. For each $U \subset Y \times Y$ we write

$$W_{X,Y}(U) \equiv \{(f,g) \in Y^X \times Y^X : \forall_{x \in X} ((f(x), g(x)) \in U)\}.$$

We show that the set

$$W \equiv \{S \subset Y^X \times Y^X : \exists_{U \in \mathcal{U}} (W_{X,Y}(U) \subset S)\} \tag{3.27}$$

is a pre-uniform structure on Y^X.

For each $U \in \mathcal{U}$, the set $W_{X,Y}(U)^{-1}$, which equals $W_{X,Y}(U^{-1})$, belongs to $\mathcal{W}$ (remember, we are taking $\mathcal{U}$ as a *uniform* structure); so if $W \in \mathcal{W}$, then $W^{-1} \in \mathcal{W}$. In view of the easily proved identities

$$W_{X,Y}(U \cap V) = W_{X,Y}(U) \cap W_{X,Y}(V),$$
$$W_{X,Y}(U) \circ W_{X,Y}(V) = W_{X,Y}(U \circ V),$$

it is straightforward to prove that $\mathcal{W}$ is a filter on $Y^X \times Y^X$ and that if $W \in \mathcal{W}$, then there exists $V \in \mathcal{W}$ with $V^2 \subset W$. Also, for $f,g \in Y^X$ we see from **U2** in Y that

$$f = g \Leftrightarrow \forall_{x \in X} (f(x) = g(x))$$
$$\Leftrightarrow \forall_{x \in X} \forall_{U \in \mathcal{U}} ((f(x), g(x)) \in U)$$
$$\Leftrightarrow \forall_{U \in \mathcal{U}} \forall_{x \in X} ((f(x), g(x)) \in U)$$
$$\Leftrightarrow \forall_{U \in \mathcal{U}} ((f,g) \in W_{X,Y}(U)).$$

It follows that $\mathcal{W}$ satisfies **U2**. This completes the proof that $\mathcal{W}$ is a pre-uniform structure on Y^X.

Note that the inequality induced by this pre-uniform structure, an inequality we denote by $\neq_\mathcal{W}$, is not the same as the standard inequality on Y^X. Indeed, as the reader is invited to show, the following statement is equivalent to Markov's principle: the standard inequality on $\{0,1\}^{\mathbf{N}}$ coincides with the inequality $\neq_\mathcal{W}$ associated with the uniform structure generated by $\{(0,0),(1,1)\}$ on $\{0,1\}$.

The pre-uniform structure $\mathcal{W}$ gives rise to a topology $\tau_{\mathcal{W}}$ on Y^X in which the sets

$$W_{X,Y}(U)[f] \equiv \{g \in Y^X : (f,g) \in W_{X,Y}(U)\} \quad (U \in \mathcal{U})$$

form a base of neighbourhoods of $f \in Y^X$. A net $(f_n)_{n \in \mathfrak{D}}$ converges to f in this topology if and only if for each $U \in \mathcal{U}$ there exists $N \in \mathfrak{D}$ such that $(f, f_n) \in W_{X,Y}(U)$—that is,

$$\forall_{x \in X} ((f(x), f_n(x)) \in U)$$

—for all $n \succcurlyeq N$. In other words, convergence with respect to $\tau_{\mathcal{W}}$ is just uniform convergence, as defined earlier. For that reason we sometimes refer to $\tau_{\mathcal{W}}$ as the **topology of uniform convergence**.

We set down here a lemma that will come into play a little later on.

Lemma 3.4.11 *Let X be an inhabited set, $(Y,\mathcal{U})$ a uniform space, and U, V entourages of Y such that $Y \times Y = U \cup {\sim}V$. Then*

$${\sim}{\sim}W_{X,Y}(V)[f] \subset W_{X,Y}(U)[f]$$

for each $f \in Y^X$.

Proof. Let $f \in Y^X$ and $g \in {\sim}{\sim}W_{X,Y}(V)[f]$. Given $x \in X$, suppose that $(f(x), g(x)) \in {\sim}V$. Then for each $h \in W_{X,Y}(V)[f]$, since $(f(x), h(x)) \in V$, we have $g(x) \neq h(x)$; whence $g \neq h$. Thus $g \in {\sim}W_{X,Y}(V)[f]$, a contradiction. Hence $(f(x), g(x)) \notin {\sim}V$ and therefore $(f(x), g(x)) \in U$. Since $x \in X$ is arbitrary, the desired conclusion now follows. $\blacksquare$

Corresponding to the pre-uniform structure $\mathcal{W}$, there is a symmetric relation $\bowtie_{\mathcal{W}}$ defined on subsets of Y^X by

$$S \bowtie_{\mathcal{W}} T \Leftrightarrow \exists_{U \in \mathcal{U}} (S \times T \subset \neg W_{X,Y}(U)).$$

Note that for $f \in Y^X$ we have

$$f \bowtie_{\mathcal{W}} T \Leftrightarrow \exists_{U \in \mathcal{U}} (W_{X,Y}(U)[f] \subset \neg T), \tag{3.28}$$

where, as one would expect, we write $f \bowtie_{\mathcal{W}} T$ as shorthand for $\{f\} \bowtie_{\mathcal{W}} T$. We define notions of complement corresponding to $\bowtie_{\mathcal{W}}$ in the obvious ways: for each $A \subset Y^X$,

$$\begin{aligned}
{\sim}_{\mathcal{W}} A &\equiv \{f \in Y^X : \forall_{g \in A} (f \neq_{\mathcal{W}} g)\}, \\
-_{\mathcal{W}} A &\equiv \{f \in Y^X : f \bowtie_{\mathcal{W}} A\}.
\end{aligned}$$

It turns out that $\bowtie_{\mathcal{W}}$ is more than just a pre-apartness:

Proposition 3.4.12 *Let X be an inhabited set, and $(Y,\mathcal{U})$ a uniform space. Then $\bowtie_{\mathcal{W}}$, taken with ${\sim}_{\mathcal{W}}$ and $-_{\mathcal{W}}$, satisfies **B1–B3** and **A4**$_{ss}$.*

Proof. Since

$$Y^X \times \varnothing = \varnothing \subset \neg \left(Y^X \times Y^X \right)$$

and

$$Y^X \times Y^X = W_X(Y \times Y) \in W,$$

axiom **B1** holds. To deal with **B2**, let $f \in Y^X$ and $A \subset Y^X$ be such that $f \in -_W A$. Then there exists $U \in \mathcal{U}$ such that for all $g \in A$, $(f,g) \in \neg W_{X,Y}(U)$ and therefore $f \neq_W g$ in Y^X.

Next let $A \bowtie_W (B \cup C)$ and choose $U \in \mathcal{U}$ such that $A \times (B \cup C) \subset \neg W_{X,Y}(U)$. Then, trivially, $A \times B \subset \neg W_{X,Y}(U)$ and $A \times C \subset \neg W_{X,Y}(U)$, so $A \bowtie_W B$ and $A \bowtie_W C$. Suppose, conversely, that $A \bowtie_W B$ and $A \bowtie_W C$. Then there exist $U, V \in \mathcal{U}$ such that $A \times B \subset \neg W_{X,Y}(U)$ and $A \times C \subset \neg W_{X,Y}(V)$; whence

$$A \times (B \cup C) \subset \neg W_{X,Y}(U) \cup \neg W_{X,Y}(V)$$
$$\subset \neg (W_{X,Y}(U) \cap W_{X,Y}(V)) = \neg W_{X,Y}(U \cap V)$$

and therefore $A \bowtie_W (B \cup C)$. Since the relation $\bowtie_W$ is symmetric, this completes the verification of **B3**.

To deal with **A4$_{ss}$**, let $A \bowtie_W B$ and $-_W B \subset \neg C$. Choose $U \in \mathcal{U}$ such that $A \times B \subset \neg W_{X,Y}(U)$, and then $V \in \mathcal{U}$ such that $V^2 \subset U$. Consider $f \in A$ and $h \in C$. We show that

$$(f, h) \in \neg W_{X,Y}(V). \tag{3.29}$$

Suppose that $(f, h) \in W_{X,Y}(V)$. For each $g \in B$, if $(h, g) \in W_{X,Y}(V)$, then

$$(f, g) \in W_{X,Y}(V) \circ W_{X,Y}(V) = W_{X,Y}(V \circ V) \subset W_{X,Y}(U),$$

which contradicts our choice of U. Hence $\{h\} \times B \subset \neg W_{X,Y}(V)$, and therefore $h \in -_W B \subset \neg C$, a further contradiction, from which (3.29) follows. Since f, h are arbitrary, we conclude that $A \times C \subset \neg W_{X,Y}(V)$ and hence that $A \bowtie_W C$. This completes the verification of **A4$_{ss}$**. ∎

The same Brouwerian example as we used on page 111 earlier in this section can easily be adapted to show that for the space $\{0,1\}^{\mathbf{N}}$, if the pre-apartness $\bowtie_W$ satisfies **A5**, then we can derive **WLPO**.

When Y is a uniform space, what, if any, are the connections between the topologies τ_p and τ_W, the pre-apartness relations $\bowtie_{YX}$ and $\bowtie_W$, and notions of convergence in Y^X?

Proposition 3.4.13 Let X be an inhabited set, and $(Y, \mathcal{U})$ a uniform space. Then uniform convergence in Y^X is equivalent to $\bowtie_W$-convergence.

Proof. Let the net $(f_n)_{n \in \mathfrak{D}}$ converge uniformly to f in Y^X, let $S \subset Y^X$, and let $f \in -_W S$. Then there exists $U \in \mathcal{U}$ such that $\{f\} \times S \subset \neg W_{X,Y}(U)$. Choose a symmetric $V \in \mathcal{U}$ such that $V^2 \subset U$. Then there exists $N \in \mathfrak{D}$

such that $(f, f_n) \in W_{X,Y}(V)$ for all $n \succeq N$. For such n, if $g \in S$ and $(f_n, g) \in W_{X,Y}(V)$, then

$$(f, g) \in W_{X,Y}(V) \circ W_{X,Y}(V) \subset W_{X,Y}(U),$$

a contradiction of our choice of U. Hence $(f_n, g) \in \neg W_{X,Y}(V)$. It follows that for all $n \succeq N$ we have $\{f_n\} \times S \subset \neg W_{X,Y}(V)$ and therefore $f_n \in -_W S$. Thus uniform convergence implies $\bowtie_W$-convergence.

Conversely, suppose that the net $(f_n)_{n \in \mathfrak{D}}$ is $\bowtie_W$-convergent to f in Y^X. Fix $U \in \mathcal{U}$ and choose $V \in \mathcal{U}$ such that $X \times X = U \cup \neg V$. Define

$$S \equiv \{f_n : n \in \mathfrak{D} \wedge (f, f_n) \in \neg W_{X,Y}(V)\}. \tag{3.30}$$

Then $\{f\} \times S \subset \neg W_{X,Y}(V)$, so $f \in -_W S$. Hence there exists N such that $f_n \in -_W S$ for all $n \succeq N$. It follows that for all such n we have $(f, f_n) \notin \neg W_{X,Y}(V)$. Let $n \succeq N$ and $x \in X$. If $(f(x), f_n(x)) \in \neg V$, then $(f, f_n) \in \neg W_{X,Y}(V)$, a contradiction; hence $(f(x), f_n(x)) \notin \neg V$ and therefore $(f(x), f_n(x)) \in U$. Since $x \in X$ and $U \in \mathcal{U}$ are arbitrary, it follows that $(f_n)_{n \in \mathfrak{D}}$ converges uniformly to f. ∎

A natural classical approach to proving the second half of Proposition 3.4.13 for sequences of functions goes as follows. Let $(f_n)_{n \geq 1}$ be a sequence that is $\bowtie_W$-convergent to f in Y^X. Fix $U \in \mathcal{U}$ and suppose that

$$\neg \exists_N \forall_{n \geq N} ((f, f_n) \in W_{X,Y}(U)).$$

Then there exists a strictly increasing sequence $(n_k)_{k \geq 1}$ such that $(f, f_{n_k}) \in \neg W_{X,Y}(U)$ for all k. It follows that

$$\{f\} \times \{f_{n_k} : k \geq 1\} \subset \neg W_{X,Y}(U)$$

and hence that

$$f \in -_W \{f_{n_k} : k \geq 1\}.$$

Now pick N such that

$$f_n \in -_W \{f_{n_k} : k \geq 1\} \quad (n \geq N).$$

Choosing j such that $n_j > N$, we see that $f_{n_j} \bowtie_W \{f_{n_j}\}$ and therefore $f_{n_j} \neq_W f_{n_j}$. This contradiction ensures that

$$\exists_N \forall_{n \geq N} ((f, f_n) \in W_{X,Y}(U)).$$

Since $U \in \mathcal{U}$ is arbitrary, we conclude that $(f_n)_{n \geq 1}$ converges uniformly to f.

This indirect proof contrasts sharply with our constructive one, in which, given $U \in \mathcal{U}$ and considering the set S introduced at (3.30), we are able to produce the desired index N such that $(f, f_n) \in W_{X,Y}(U)$ for all $n \succeq N$.

Proposition 3.4.14 Let X be an inhabited set, and $(Y,\mathcal{U})$ a uniform space. Then the topology τ_W of uniform convergence on Y^X is finer than the topology τ_p of proximal convergence.

Proof. Let $f \in Y^X$, and let A_i, B_i $(1 \leqslant i \leqslant m)$ be subsets of X,Y respectively, such that $f \in \bigcap_{i=1}^{m} U_{A_i,B_i}$. Then for each i we have $f(A_i) \bowtie B_i$ in Y, so there exists $U_i \in \mathcal{U}$ such that $f(A_i) \times B_i \subset \neg U_i$. Let

$$V \equiv \bigcap_{i=1}^{m} U_i,$$

which belongs to $\mathcal{U}$. Then

$$f(A_i) \times B_i \subset \neg V \quad (1 \leqslant i \leqslant m).$$

Pick $V_2 \in \mathcal{U}$ such that (V, V_2) is a 2-chain, and consider any $g \in W_{X,Y}(V_2)[f]$. Given $x \in A_i$ and $y \in B_i$, suppose that $(g(x), y) \in V_2$. Since $(f, g) \in W_{X,Y}(V_2)$, we have $(f(x), g(x)) \in V_2$ and therefore $(f(x), y) \in V_2^2 \subset V$, a contradiction. It follows that $g(A_i) \times B_i \subset \neg V_2$, so $g(A_i) \bowtie B_i$ and therefore $g \in U_{A_i,B_i}$. Hence

$$W_{X,Y}(V_2)[f] \subset \bigcap_{i=1}^{m} U_{A_i,B_i}.$$

It follows that every neighbourhood of f in (Y^X, τ_p) contains some neighbourhood of f in (Y^X, τ_W). ∎

We can use Proposition 3.4.14 to prove Proposition 3.4.3, as follows. Let X be an inhabited set, $(Y,\mathcal{U})$ a uniform space, and $(f_n)_{n \in \mathfrak{D}}$ a net in Y^X that converges uniformly to $f \in Y^X$. Let $A \subset X$ and $B \subset Y$ be such that $f(A) \bowtie B$. Then $f \in U_{A,B}$. By Proposition 3.4.14, there exists $U \in \mathcal{U}$ such that $W_{X,Y}(U)[f] \subset U_{A,B}$. Choose $N \in \mathfrak{D}$ such that $(f, f_n) \in U$, and therefore $f_n \in W_{X,Y}(U)[f]$, for all $n \succcurlyeq N$. For all such n we have $f_n \in U_{A,B}$ and therefore $f_n(A) \bowtie f(B)$.

Proposition 3.4.15 Let X be an inhabited set, and $(Y,\mathcal{U})$ a totally bounded uniform space. Then the topologies τ_p and τ_W coincide.

Proof. In view of Proposition 3.4.14, it is enough to prove that τ_p is finer than τ_W. In turn, given $f \in Y^X$ and an entourage U of Y, we need only prove that $W_{X,Y}(U)[f]$ contains a τ_p-neighbourhood of f. To that end, construct a 4-chain (U, U_1, U_2, U_3) of entourages of Y such that U_1, U_3 are symmetric, $U_1^3 \subset U$, and $U_3^3 \subset U_2$. Choose $y_1, \ldots, y_m$ in X such that $Y = Y_1 \cup \cdots \cup Y_m$, where

$$Y_i \equiv U_3[y_i].$$

For $1 \leqslant i, j \leqslant m$ construct c_{ij} such that

$$c_{ij} = 0 \Rightarrow (y_i, y_j) \in U_1,$$
$$c_{ij} = 1 \Rightarrow (y_i, y_j) \in \neg U_2.$$

We prove that

(i) if $c_{ij} = 0$, then $Y_i \times Y_j \subset U$, and

(ii) if $c_{ij} = 1$, then $Y_i \bowtie Y_j$.

Suppose that $c_{ij} = 0$, and let $y \in Y_i, y' \in Y_j$. Then $(y, y_i) \in U_3^{-1} = U_3$ and $(y_j, y') \in U_3$, so

$$(y, y') \in U_3 \circ U_1 \circ U_3 \subset U_1^3 \subset U.$$

This proves (i). For (ii), let $c_{ij} = 1$, consider $y \in Y_i$ and $y' \in Y_j$, and suppose that $(y, y') \in U_3$. Then $(y_i, y) \in U_3$ and $(y', y_j) \in U_3^{-1} = U_3$, so $(y_i, y_j) \in U_3^3 \subset U_2$, which is absurd since $c_{ij} = 1$. Hence $Y_i \times Y_j \subset \neg U_3$ and therefore $Y_i \bowtie Y_j$. This completes the proof of (ii).

If $c_{ij} = 0$ for all i and j, then for all $g \in Y^X$ and all $x \in X$ we have $(f(x), g(x)) \in U$, by (i); whence $f \in Y^X \subset W_{X,Y}(U)[f]$. Thus we may assume that there exist i, j with $c_{ij} = 1$. Writing $X_i \equiv f^{-1}(Y_i)$, we see from (ii) that

$$S \equiv \bigcap \{U_{X_i, Y_j} : 1 \leqslant i, j \leqslant m \wedge c_{ij} = 1\}$$

is a τ_p-neighbourhood of f. Consider any g in this neighbourhood and any $x \in X$. Choose i such that $f(x) \in Y_i$ and therefore $x \in X_i$. Choose also j such that $g(x) \in Y_j$. If $c_{ij} = 1$, then our choice of g ensures that $g(X_i) \bowtie Y_j$ and therefore $g(x) \notin Y_j$, a contradiction. Thus $c_{ij} = 0$, so $(f(x), g(x)) \in U$, by (i). Since g and x are arbitrary, we conclude that the τ_p-neighbourhood S of f is a subset of $W_{X,Y}(U)[f]$. ∎

Corollary 3.4.16 Let X be an inhabited set, and $(Y, \mathcal{U})$ a totally bounded uniform space. Then proximal convergence and uniform convergence in Y^X are equivalent.

Proof. This follows from Propositions 3.4.9 and 3.4.15, since uniform convergence is equivalent to convergence with respect to the topology τ_W. ∎

Uniform spaces for which the topologies of proximal and uniform convergence coincide are worth a second glance.

Proposition 3.4.17 Let X be an inhabited set, and $(Y, \mathcal{U})$ a uniform space such that the topologies τ_p and τ_W coincide. Then (Y^X, τ_p) is a topologically consistent pre-apartness space.

Proof. Let A be a τ_p-open set, and $f \in A$. Since $\tau_p \subset \tau_W$, there exists $U \in \mathcal{U}$ such that $f \in W_{X,Y}(U)[f] \subset A$. Using Proposition 3.2.6, construct $V \in \mathcal{U}$ such that $Y \times Y = U \cup {\sim} V$. By our hypothesis and Lemma 3.4.11, there exist subsets $A_1, \ldots, A_m$ of X and $B_1, \ldots, B_m$ of Y such that

$$f \in \bigcap_{i=1}^{m} U_{A_i, B_i} \subset W_{X,Y}(V)[f] \subset {\sim}{\sim} W_{X,Y}(V)[f] \subset W_{X,Y}(U)[f].$$

Denoting the apartness complement relative to $\bowtie_{Yx}$ by $-_{Yx}$, we obtain

$$f \in -_{Yx} {\sim} W_{X,Y}(V)[f] \subset A.$$

Thus A is a neighbourhood of f in the apartness topology induced by $\bowtie_{Yx}$. Since $f \in A$ is arbitrary, we conclude that A is nearly open relative to $\bowtie_{Yx}$. ∎

Corollary 3.4.18 *Let X be an inhabited set, and $(Y, \mathcal{U})$ a totally bounded uniform space. Then (Y^X, τ_p) is a topologically consistent pre-apartness space.*

Proof. This is a simple consequence of Propositions 3.4.15 and 3.4.17. ∎

Proposition 3.4.19 *Let X be an inhabited set, and $(Y, \mathcal{U})$ a uniform space. Then*
$$\forall_{f \in Y^X} \forall_{S \subset Y^X} (f \bowtie_{Yx} S \Rightarrow f \bowtie_W S).$$

Proof. Let $f \bowtie_{Yx} S$, and pick a τ_p-open set $A \subset Y^X$ such that $f \in A \subset {\sim} S$. By Proposition 3.4.14, there exists $U \in \mathcal{U}$ such that $W_{X,Y}(U)[f] \subset A$, so $W_{X,Y}(U)[f] \subset {\sim} S$. It follows that if $g \in S$, then $(f, g) \notin W_{X,Y}(U)$. Thus $\{f\} \times S \subset \neg W_{X,Y}(U)$ and therefore $f \bowtie_W S$. ∎

Under the hypotheses of Proposition 3.4.19, if the converse implication

$$\forall_{f \in Y^X} \forall_{S \subset Y^X} (f \bowtie_W S \Rightarrow f \bowtie_{Yx} S) \tag{3.31}$$

holds, then the point-set relations $\bowtie_{Yx}$ and $\bowtie_W$ coincide and our various notions of convergence in Y^X coalesce:

Proposition 3.4.20 *Let X be an inhabited set, and $(Y, \mathcal{U})$ a uniform space with the property (3.31). Then $\bowtie_{Yx}$-convergence, uniform convergence, proximal convergence, τ_p-convergence, and τ_W-convergence are all equivalent.*

Proof. We already know (from Proposition 3.4.3) that uniform convergence implies proximal convergence, and (from Corollary 3.4.10) that proximal convergence implies $\bowtie_{Yx}$-convergence. On the other hand, by Proposition 3.4.13, $\bowtie_W$-convergence is equivalent to uniform convergence. Now, by (3.31) and Proposition 3.4.19, $\bowtie_{Yx} = \bowtie_W$, so $\bowtie_{Yx}$-convergence is equivalent to uniform convergence. It remains to invoke Proposition 3.4.9 and the observation, made earlier, that τ_W-convergence is just uniform convergence. ∎

Corollary 3.4.21 *Under the hypotheses of* Proposition 3.4.20, (Y^X, τ_p) *is a topologically consistent pre-apartness space.*

Proof. This follows from Proposition 3.4.20 and Corollary 3.4.10. ∎

This corollary fits with what we have learned about Y^X when Y is a totally bounded uniform space: in that case, Corollary 3.4.18 tells us that (Y^X, τ_p) is topologically consistent, and Proposition 3.4.15 that proximal convergence—equivalent (by Proposition 3.4.9) to τ_p-convergence—implies τ_W-convergence, which is equivalent to uniform convergence.

Nachman [73] has shown classically, in the context of proximity spaces, that proximal convergence need not imply uniform convergence. Consequently, when Y is a uniform space, we have no guarantee that the space (Y^X, τ_p) satisfies (3.31).

A diagram may help to clarify the relations between various types of convergence in Y^X when Y is a uniform space:

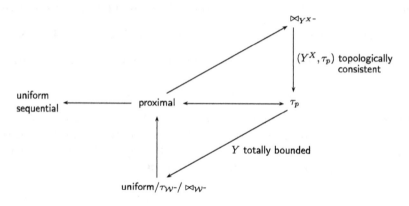

3.5   Totally Cauchy Nets

Having discussed convergence at some length in Chapter 2 and the preceding section of this chapter, we next consider types of Cauchy net in the context of pre-apartness and uniform spaces.

In the classical theory of proximity [35], a sequence $s \equiv (x_n)_{n \geqslant 1}$ in a proximity space X is called a *Cauchy sequence* if for all infinite sets A and B of positive integers, the sets $s(A)$ and $s(B)$ are near each other, where, for example,

$$s(A) \equiv \{x_n : n \in A\}.$$

We call such a sequence a **CHN-Cauchy sequence**. Bearing in mind that the canonical proximity relation between subsets S, T of a metric space (X, ρ) is defined by

$$S \text{ is near } T \text{ if and only if } \forall_{\varepsilon > 0} \exists_{s \in S} \exists_{t \in T} \ (\rho\,(s,t) < \varepsilon)\,,$$

we can easily show that in a metric space (X, ρ) a **metrically Cauchy se-quence**—that is, a sequence $(x_n)_{n \geqslant 1}$ such that

$$\forall_{\varepsilon > 0} \exists_N \forall_{m,n \geqslant N} \ (\rho(x_m, x_n) < \varepsilon)$$

—is a CHN-Cauchy sequence. The converse is proved in [35] by an indirect, and therefore nonconstructive, argument. The following Brouwerian example shows that we cannot prove that converse constructively.

Let $s \equiv (x_n)_{n \geqslant 1}$ be an increasing binary sequence, and A, B infinite sets of positive integers. Then $s(A)$ and $s(B)$ intersect and so are near each other. (To see this, pick $m \in A$ and $n \in B$; we may assume that $s(m) \neq s(n)$, so one of these, say $s(m)$, equals 1. Since B is infinite, there exists $n' \in B$ such that $n' > m$; then $s(n') = 1 = s(m)$.) However, if s is a metrically Cauchy sequence, then there exists N such that $|x_m - x_n| < 1$ for all $m, n \geqslant N$; by testing $x_1, \ldots, x_N$, we can determine that either $x_n = 0$ for all n or else there exists $n \leqslant N$ such that $x_n = 1$. Thus the statement

> Every binary CHN-Cauchy sequence is a metrically Cauchy se-quence

entails **LPO**.

Now let X be any pre-apartness space. We want to define a Cauchy property for sequences that is classically equivalent to the CHN-Cauchy property, and that in a metric space is constructively equivalent to metric Cauchyness. The following definition introduces such a property in the more general setting of nets.

A net $s \equiv (x_n)_{n \in \mathfrak{D}}$ in X is called a **totally Cauchy net** if for all subsets A, B of $\mathfrak{D}$ with $s(A) \bowtie s(B)$, there exists $N \in \mathfrak{D}$ such that

$$\neg\,(\exists_{m \in A} \,(m \succcurlyeq N) \land \exists_{n \in B} \,(n \succcurlyeq N))\,. \qquad (3.32)$$

Although, in the light of experience, we might reasonably expect that conver-gent nets are totally Cauchy, life is not that simple. To see this, we introduce a notion that will re-appear on a number of occasions: we define a pre-apartness space X to be **weakly symmetrically separated** if, for all $S, T \subset X$,

$$S \bowtie T \Rightarrow \forall_{x \in X} \, \exists_{U \subset X} \,(x \in -U \land \neg\,(S - U \neq \varnothing \land T - U \neq \varnothing))\,. \qquad (3.33)$$

If **B7** holds (see page 86), then the space is weakly symmetrically separated: for in that case, if $S \bowtie T$ in X and $x \in X$, we can find $U \subset X$ with $x \in -U$ such that if $S - U$ is inhabited, then $\neg U \bowtie T$; whence

$$-U \subset \neg U \subset -T \subset \neg T.$$

It follows from this and Proposition 3.2.26 that every uniform apartness space is weakly symmetrically separated.

With classical logic we can prove that every symmetric set-set pre-apartness space X satisfying **A4**$_s$ is weakly symmetrically separated: let $S \bowtie T$ in X, and suppose that there exists $x \in X$ such that for all U with $x \in -U$, both $S - U$ and $T - U$ are inhabited; then $x \in \overline{S} \cap \overline{T}$ and so $\neg (\overline{S} \bowtie \overline{T})$, which (since the reverse Kolmogorov property automatically holds in the presence of the law of excluded middle) contradicts Corollary 3.1.19.

Proposition 3.5.1 *The following are equivalent conditions on a pre-apartness space $(X, \bowtie)$:*

(i) *Every convergent net in X is a totally Cauchy net.*

(ii) *X is weakly symmetrically separated.*

Proof. Assuming (i), let $S \bowtie T$ and $x \in X$, and consider the index set $\mathfrak{D}_x$ and the neighbourhood net $\mathcal{N}_x$, both defined on page 41. Defining

$$A \equiv \{n \in \mathfrak{D}_x : \mathcal{N}_x(n) \in S\},$$
$$B \equiv \{n \in \mathfrak{D}_x : \mathcal{N}_x(n) \in T\},$$

we have $\mathcal{N}_x(A) \bowtie \mathcal{N}_x(B)$. Since $\mathcal{N}_x$ converges to x, we see from (i) that there exists $N \equiv (y_0, U_0) \in \mathfrak{D}_x$ such that (3.32) holds. Suppose there exist $y \in S - U_0$ and $z \in T - U_0$. Then $(y, U_0) \in \mathfrak{D}_x$ and $\mathcal{N}_x(y, U_0) = y \in S$, so $(y, U_0) \in A$; also, $(y, U_0) \succcurlyeq (y_0, U_0) = N$. Likewise, $(z, U_0) \in B$ and $(z, U_0) \succcurlyeq N$. This contradicts (3.32). Hence

$$\neg (S - U_0 \neq \varnothing \wedge T - U_0 \neq \varnothing)$$

and X is weakly symmetrically separated.

Now assume (ii), let $s \equiv (x_n)_{n \in \mathfrak{D}}$ be a net converging to an element x in X, and let A, B be subsets of $\mathfrak{D}$ such that $s(A) \bowtie s(B)$. By our hypothesis (ii), there exists $U \subset X$ such that

$$x \in -U \wedge \neg (s(A) - U \neq \varnothing \wedge s(B) - U \neq \varnothing).$$

Choose N in $\mathfrak{D}$ such that $x_n \in -U$ for all $n \succcurlyeq N$. Suppose that for some $m, n \succcurlyeq N$ we have $m \in A$ and $n \in B$. Then $x_m \in s(A) - U$ and $x_n \in s(B) - U$, which is absurd. Hence

$$\neg (\exists_{m \succcurlyeq N} (m \in A) \wedge \exists_{n \succcurlyeq N} (n \in B)),$$

and therefore the convergent net s is totally Cauchy. ∎

Corollary 3.5.2 *If X satisfies **B7**, then every convergent net in X is totally Cauchy.*

Proof. Since condition **B7** implies (3.33), we can apply Proposition 3.5.1. ∎

Corollary 3.5.3 *Every convergent net in a uniform apartness space is totally Cauchy.*

Proof. This follows from Proposition 3.2.26 and Corollary 3.5.2. ∎

A net $s \equiv (x_n)_{n \in \mathfrak{D}}$ in a quasi-uniform space $(X, \mathcal{U})$ is called a **uniformly Cauchy net** if

$$\forall_{U \in \mathcal{U}} \exists_N \forall_{m,n \succcurlyeq N} ((x_m, x_n) \in U).$$

It is an exercise to show that every convergent net in a uniform space is uniformly Cauchy.

Proposition 3.5.4 *Every uniformly Cauchy net in a quasi-uniform space is totally Cauchy.*

Proof. Let $s \equiv (x_n)_{n \in \mathfrak{D}}$ be a uniformly Cauchy net in a quasi-uniform space $(X, \mathcal{U})$, and let $A, B \subset \mathfrak{D}$ be such that $s(A) \bowtie s(B)$. Then there exists an entourage U such that $s(A) \times s(B) \subset \neg U$. Since s is a uniformly Cauchy net, there exists $N \in \mathfrak{D}$ such that $(x_m, x_n) \in U$ for all $m, n \succcurlyeq N$. If $p, q \succcurlyeq N$ and $p \in A$ and $q \in B$, then $(x_p, x_q) \in (s(A) \times s(B)) \cap U$, which is absurd. It follows that s is totally Cauchy. ∎

Here is a partial converse to Proposition 3.5.4.

Theorem 3.5.5 *A totally Cauchy net in a totally bounded uniform space is uniformly Cauchy.*

Proof. Let $s \equiv (x_n)_{n \in \mathfrak{D}}$ be a totally Cauchy net in the totally bounded uniform space $(X, \mathcal{U})$. Given an entourage U_1 of X, construct a 5-chain $(U_1, \ldots, U_5)$ of entourages such that U_5 is symmetric, $U_2^3 \subset U_1, U_3^3 \subset U_2$, and $U_5^3 \subset U_4$. Choose $a_1, \ldots, a_m$ in X such that $X = \bigcup_{i=1}^{m} U_5[a_i]$. For all i, j $(1 \leqslant i, j \leqslant m)$ compute $c_{ij} \in \{0, 1\}$ such that

$$c_{ij} = 0 \Rightarrow (a_i, a_j) \in U_3,$$
$$c_{ij} = 1 \Rightarrow (a_i, a_j) \in \neg U_4.$$

Also, define

$$A_i \equiv \{n \in \mathfrak{D} : (x_n, a_i) \in U_5\}.$$

We prove the following:

(i) if $c_{ij} = 0$, then $U_5[a_i] \times U_5[a_j] \subset U_2$;

(ii) if $c_{ij} = 1$, then there exists $n_{ij} \in \mathfrak{D}$ such that

$$\neg \left(\exists_{m \in A_i} (m \succcurlyeq n_{ij}) \wedge \exists_{n \in A_j} (n \succcurlyeq n_{ij}) \right). \tag{3.34}$$

To prove (i) is straightforward: if $c_{ij} = 0$, $x \in U_5[a_i]$, and $y \in U_5[a_j]$, then

$$(x, y) \in U_5^{-1} \circ U_3 \circ U_5 \subset U_3^3 \subset U_2.$$

For (ii), consider i, j with $c_{ij} = 1$. Given $p \in A_i$ and $q \in A_j$, suppose that $(x_p, x_q) \in U_5$. Then

$$(a_i, a_j) \in U_5^{-1} \circ U_5 \circ U_5 = U_5^3 \subset U_4,$$

which is absurd since $c_{ij} = 1$. It follows that $(x_p, x_q) \in \neg U_5$. Thus $s(A_i) \times s(A_j) \subset \neg U_5$ and therefore $s(A_i) \bowtie_{\mathcal{U}} s(A_j)$. Since s is totally Cauchy, there exists n_{ij} such that (3.34) holds. This completes the proof of (ii).

If $c_{ij} = 0$ for all i and j, then it follows from (i) that $(x_m, x_n) \in U_2 \subset U_1$ for all $m, n \in \mathfrak{D}$. Thus we may assume that $c_{ij} = 1$ for some i and j. Choose $N \in \mathfrak{D}$ such that $N \succcurlyeq n_{ij}$ for all i, j with $c_{ij} = 1$. Let $m, n \succcurlyeq N$, and pick i, j such that $m \in A_i$ and $n \in A_j$. In view of (ii), we cannot have $c_{ij} = 1$, so $c_{ij} = 0$ and hence $(a_i, a_j) \in U_2$. On the other hand, (x_m, a_i) and (x_n, a_j) belong to U_5, which is a symmetric subset of U_2, so

$$(x_m, x_n) \in U_5 \circ U_2 \circ U_5^{-1} \subset U_2^3 \subset U_1$$

for all $m, n \succcurlyeq N$. Thus s is uniformly Cauchy. ∎

Corollary 3.5.6 *A totally bounded, totally Cauchy net in a uniform space is uniformly Cauchy.*

Proof. Apply the preceding theorem in the uniform subspace comprising all terms of a totally bounded, totally Cauchy net in the uniform space. ∎

Using a sequence of at-times-complicated lemmas, we shall produce alternative hypotheses under which a totally Cauchy net in a uniform space is uniformly Cauchy.

Lemma 3.5.7 *Let X be a uniform apartness space, and $s \equiv (x_n)_{n \in \mathfrak{D}}$ a totally Cauchy net in X. Let (U_1, U_2, U_3, U_4) be a 4-chain of entourages of X with U_3 symmetric, and let $n_0, m_0 \in \mathfrak{D}$. Then there exists n_1 such that $n_1 \succcurlyeq n_0$, $n_1 \succcurlyeq m_0$, and either $(x_{n_1}, x_{n_0}) \in \neg U_4$ or else $(x_n, x_{n_0}) \in U_1$ for all $n \succcurlyeq n_1$.*

Proof. Define

$$A \equiv \{ n : (x_n, x_{n_0}) \in U_3 \},$$
$$B \equiv \{ n : (x_n, x_{n_0}) \in \neg U_2 \}.$$

If $n \in A$, $n' \in B$, and $(x_n, x_{n'}) \in U_3$, then, since U_3 is symmetric, $(x_{n'}, x_{n_0}) \in U_3^2 \subset U_2$, a contradiction. Hence

$$\forall_{n \in A} \forall_{n' \in B} ((x_n, x_{n'}) \in \neg U_3)$$

and therefore $s(A) \bowtie s(B)$. Thus we can find $N \in \mathfrak{D}$ such that (3.32) holds. Choose $n_1 \in \mathfrak{D}$ such that $n_1 \succcurlyeq N$, $n_1 \succcurlyeq n_0$, and $n_1 \succcurlyeq m_0$. Either $(x_{n_1}, x_{n_0}) \in \neg U_4$ or else $(x_{n_1}, x_{n_0}) \in U_3$. In the latter case, $n_1 \in A$; whence for each $n \succcurlyeq n_1$ we have $n \notin B$, $(x_n, x_{n_0}) \notin \neg U_2$, and therefore $(x_n, x_{n_0}) \in U_1$. $\blacksquare$

Let $\mathfrak{D}$ be a directed set. A subset $\mathfrak{E}$ of $\mathfrak{D}$ is said to be **cofinal** (with $\mathfrak{D}$) if for each $n \in \mathfrak{D}$ there exists $m \in \mathfrak{E}$ with $m \succcurlyeq n$; in that case, $\mathfrak{E}$ itself is directed. By a **subnet** of a net $(x_n)_{n \in \mathfrak{D}}$ we mean a net $(x_n)_{n \in \mathfrak{E}}$ where $\mathfrak{E}$ is a cofinal subset of $\mathfrak{D}$.

Lemma 3.5.8 Let X be a pre-apartness space, and $s \equiv (x_n)_{n \in \mathfrak{D}}$ a net in X that converges to a limit $x \in X$. Then every subnet of s converges to x.

Proof. Let $\mathfrak{E}$ be a cofinal subset of $\mathfrak{D}$, and S a subset of X such that $x \in -S$. There exists N such that $x_n \in -S$ whenever $n \in \mathfrak{D}$ and $n \succcurlyeq N$. Since $\mathfrak{E}$ is cofinal, we can find $\nu \in \mathfrak{E}$ such that $\nu \succcurlyeq N$. For each $n \in \mathfrak{E}$ with $n \succcurlyeq \nu$ we have $n \succcurlyeq N$ and therefore $x_n \in -S$. Hence the subnet $(x_n)_{n \in \mathfrak{E}}$ converges to x in X. $\blacksquare$

Lemma 3.5.9 Let A, B be cofinal subsets of a directed set $\mathfrak{D}$. If $s \equiv (x_n)_{n \in \mathfrak{D}}$ is a totally Cauchy net in a pre-apartness space X, then $\neg (s(A) \bowtie s(B))$.

Proof. Suppose that $s(A) \bowtie s(B)$, and pick $N \in \mathfrak{D}$ such that (3.32) holds. Since A is cofinal, there exists $n \in A$ with $n \succcurlyeq N$; so for each $n \in B$ we have $n \nsucccurlyeq N$, which contradicts the cofinality of B. $\blacksquare$

Lemma 3.5.10 Let X be a uniform apartness space, $s \equiv (x_n)_{n \in \mathfrak{D}}$ a totally Cauchy net in X, and $(U_1, U_2, U_3, U_4, U_5)$ a 5-chain of symmetric entourages of X. Suppose that $\mathfrak{D}$ has a countable cofinal set $\mathfrak{E} \equiv \{m_k : k \geqslant 0\}$. Then

▷ either there exists N such that $(x_n, x_N) \in U_1$ for all $n \succcurlyeq N$

▷ or else there exist $n_1 \preccurlyeq n_2 \preccurlyeq n_3 \preccurlyeq \cdots$ in $\mathfrak{D}$ such that $n_{k+1} \succcurlyeq m_k$ and $(x_{n_{k+1}}, x_{n_k}) \in \neg U_5$ for each positive integer k.

Proof. Pick an element n_0 of $\mathfrak{D}$, let $\lambda_0 = 0$, and let both A_0 and B_0 equal the empty subset of $\mathfrak{D}$. Suppose that for some k we have constructed $\lambda_k \in \{0, 1\}$, $n_k \in \mathfrak{D}$, and subsets A_k, B_k of $\mathfrak{D}$. If $\lambda_k = 1$, we set $\lambda_{k+1} \equiv 1$, $A_{k+1} \equiv B_{k+1} \equiv \varnothing$, and $n_{k+1} \equiv n_k$. If $\lambda_k = 0$, we proceed as follows. Applying Lemma 3.5.7, we construct $n_{k+1} \in \mathfrak{D}$ such that $n_{k+1} \succcurlyeq n_k$, $n_{k+1} \succcurlyeq m_k$, and either $(x_{n_{k+1}}, x_{n_k}) \in \neg U_5$ or else $(x_n, x_{n_k}) \in U_2$ for all $n \succcurlyeq n_{k+1}$. In the first case, we set $\lambda_{k+1} \equiv 0$ and $A_{k+1} \equiv B_{k+1} \equiv \varnothing$. In the second, we set $\lambda_{k+1} \equiv 1$; we also set $A_1 \equiv B_1 \equiv \varnothing$ if $k = 0$, and $A_{k+1} \equiv \{n_k\}$, $B_{k+1} \equiv \{n_{k-1}\}$ if $k \geqslant 1$. This completes the inductive construction of $\lambda_{k+1}, n_{k+1}, A_{k+1}$, and B_{k+1}.

Now let

$$A \equiv \bigcup_{k \geqslant 1} A_k, \quad B \equiv \bigcup_{k \geqslant 1} B_k.$$

If $n \in A$ and $n' \in B$, then there exists $k \geqslant 1$ such that $\lambda_{k+1} = 1 - \lambda_k$, $n \in A_{k+1} = \{n_k\}$, $n' \in B_{k+1} = \{n_{k-1}\}$; moreover, since $\lambda_{k-1} = 0$, we have $(x_{n_k}, x_{n_{k-1}}) \in \neg U_5$. It follows that $s(A) \times s(B) \subset \neg U_5$ and therefore $s(A) \bowtie s(B)$. Thus, by total Cauchyness, there exists ν such that if $n \in A$ for some $n \succcurlyeq \nu$, then $n \not\succcurlyeq \nu$ for each $n \in B$. Since $\mathfrak{C}$ is cofinal, we can pick $K > 1$ such that $m_{K-1} \succcurlyeq \nu$. If $\lambda_{K+1} = 1$, then there exists $k \leqslant K$ such that $\lambda_{k+1} = 1 - \lambda_k$ and therefore $(x_n, x_{n_k}) \in U_2$ for all $n \succcurlyeq n_{k+1}$. In particular, $(x_{n_{k+1}}, x_{n_k}) \in U_2$; so $(x_n, x_{n_{k+1}}) \in U_2 \circ U_2^{-1} \subset U_1$ for all $n \succcurlyeq n_{k+1}$, and we obtain the first alternative in the desired conclusion of the lemma by taking $N \equiv n_{k+1}$. On the other hand, if $\lambda_{K+1} = 0$, then $\lambda_k = 0$ for all $k > K$ and therefore for all k: for, supposing that $\lambda_{k+1} = 1 - \lambda_k$ for some $k > K$, we have $A = \{n_k\}, B = \{n_{k-1}\}$, and

$$n_k \succcurlyeq n_{k-1} \succcurlyeq n_K \succcurlyeq m_{K-1} \succcurlyeq \nu,$$

which, by our choice of ν, is impossible. Thus in the case $\lambda_{K+1} = 0$, we have $n_{k+1} \succcurlyeq m_k$ and $(x_{n_{k+1}}, x_{n_k}) \in \neg U_5$ for each k, and $n_1 \preccurlyeq n_2 \preccurlyeq n_3 \preccurlyeq \cdots$. ∎

Lemma 3.5.11 *Under the hypotheses of* Lemma 3.5.10, *if the second conclusion holds, then so does* **LPO**.

Proof. Consider any increasing binary sequence $(\lambda_k)_{k \geqslant 1}$ with $\lambda_1 = 0$. If $\lambda_k = 0$, set $A_k = B_k \equiv \varnothing$; if $\lambda_k = 1 - \lambda_{k-1}$, set $A_k \equiv \{n_{k+1}\}$, $B_k \equiv \{n_k\}$, and $A_j = B_j \equiv \varnothing$ for every $j > k$. Let

$$A \equiv \bigcup_{k \geqslant 1} A_k, \quad B \equiv \bigcup_{k \geqslant 1} B_k.$$

If $n \in A$ and $n' \in B$, then $n = n_{k+1}$ and $n' = n_k$ for some k such that $\lambda_k = 1 - \lambda_{k-1}$; so

$$s(A) \times s(B) = \{(x_{n_{k+1}}, x_{n_k})\} \subset \neg U_5.$$

It follows that $s(A) \bowtie s(B)$. Thus we can choose N such that (3.32) holds. Pick K such that $m_K \succcurlyeq N$. If $\lambda_K = 1$, then there is nothing to prove. If $\lambda_K = 0$, suppose that $\lambda_k = 1 - \lambda_{k-1}$ for some $k > K$. Then $A = \{n_{k+1}\}$ and $B = \{n_k\}$, where $n_{k+1} \succcurlyeq n_k \succcurlyeq n_{K+1} \succcurlyeq m_K \succcurlyeq N$. Since $n_{k+1} \in A$ and $n_k \in B$, this contradicts our choice of N. Hence $\lambda_k = 0$ for all $k > K$ and therefore for all k. ∎

We now complete our link between totally Cauchy and uniformly Cauchy nets.

Theorem 3.5.12 *A totally Cauchy net with a countable subnet in a uniform space is uniformly Cauchy.*

Proof. Let $(X, \mathcal{U})$ be a uniform space, and let $s \equiv (x_n)_{n \in \mathfrak{D}}$ be a totally Cauchy net in X with a countable subnet $(x_{m_k})_{k \geqslant 1}$. Given an entourage U_1 in

$\mathcal{U}$, construct an 8-chain $(U_1, \ldots, U_8)$ of entourages of X such that $U_7^3 \subset U_6$ and U_k is symmetric for $k > 1$. By Lemma 3.5.10, either there exists N such that $(x_n, x_N) \in U_2$ for all $n \succcurlyeq N$, or else there exist $n_1 \preccurlyeq n_2 \preccurlyeq \cdots$ in $\mathfrak{D}$ such that $n_{k+1} \succcurlyeq m_k$ and $(x_{n_{k+1}}, x_{n_k}) \in \neg U_6$ for each k. In the first case, $(x_m, x_n) \in U_2 \circ U_2^{-1} \subset U_1$ for all $m, n \succcurlyeq N$. In the second case, which we shall rule out, we obtain **LPO** from Lemma 3.5.11. Taking

$$a_k \equiv x_{n_{k+1}}, \ b_k \equiv x_{n_k} \quad (k \geqslant 1),$$

we can therefore apply Lemma 3.3.17 to the 3-chain (U_6, U_7, U_8), to construct a strictly increasing sequence $(p_k)_{k \geqslant 1}$ of positive integers such that

$$\left(x_{n_{p_j+1}}, x_{n_{p_k}} \right) = (a_{p_j}, b_{p_k}) \in \neg U_8$$

for all j and k. The sets

$$A \equiv \{n_{p_k+1} \in \mathfrak{D} : k \geqslant 1\} \text{ and } B \equiv \{n_{p_k} \in \mathfrak{D} : k \geqslant 1\}$$

are cofinal: for, given $n \in \mathfrak{D}$ and picking k such that $m_{k-1} \succcurlyeq n$, we have $n_{p_k+1} \succcurlyeq n_{p_k} \succcurlyeq n_k \succcurlyeq m_{k-1} \succcurlyeq n$, where $n_{p_k+1} \in A$ and $n_{p_k} \in B$. But $s(A) \times s(B) \subset \neg U_8$, so $s(A) \bowtie s(B)$, which contradicts Lemma 3.5.9. This completes the proof. ∎

Proposition 3.5.13 *A sequence in a uniform space is totally Cauchy if and only if it is uniformly Cauchy.*

Proof. This follows from Proposition 3.5.4 and Theorem 3.5.12 (since $\mathbf{N}^+$ is a countable cofinal subset of itself). ∎

We now establish a number of basic properties of totally Cauchy nets, and look at spaces with a related property of total completeness.

Proposition 3.5.14 *If $f : X \to Y$ is a strongly continuous map between pre-apartness spaces, and $s \equiv (x_n)_{n \in \mathfrak{D}}$ is a totally Cauchy net in X, then $(f(x_n))_{n \in \mathfrak{D}}$ is a totally Cauchy net in Y.*

Proof. Let A, B be subsets of $\mathfrak{D}$ such that $f(s(A)) \bowtie f(s(B))$ in Y. By the strong continuity of f, $s(A) \bowtie s(B)$ in X. The desired conclusion follows immediately, since s is a totally Cauchy net in X. ∎

A pre-apartness space X is said to have the **nested neighbourhoods property** if

$$x \in -U \Rightarrow \exists_{V \subset X} (x \in -V \wedge \neg V \bowtie U).$$

This property, which clearly implies the weak nested neighbourhoods property, is a trivial consequence of **B7** and hence, by Proposition 3.2.26, holds in every uniform space.

Proposition 3.5.15 *Let X be a pre-apartness space with the nested neigh-bourhoods property, and let $s \equiv (x_n)_{n \in \mathfrak{D}}$ be a totally Cauchy net that contains a subnet converging to a limit x in X. Then s converges to x.*

Proof. Let $(x_i)_{i \in I}$ be a subnet of s converging to x in X, and let $x \in -U$ in X. The nested neighbourhoods property ensures that there exists $V \subset X$ such that $x \in -V$ and $\neg V \bowtie U$; again using the nested neighbourhoods property, we can find $W \subset X$ such that $x \in -W$ and $\neg W \bowtie V$. Let

$$A \equiv \{n \in \mathfrak{D} : x_n \in -W\},$$
$$B \equiv \{n \in \mathfrak{D} : x_n \in V\}.$$

Since $-W \subset \neg W$, we have $s(A) \bowtie s(B)$; whence there exists $n_0 \in \mathfrak{D}$ such that if $n \succcurlyeq n_0$ for some $n \in A$, then

$$B \subset \neg \{n : n \succcurlyeq n_0\}. \tag{3.35}$$

On the other hand, since $(x_i)_{i \in I}$ converges to x, there exists $i_0 \in I$ such that $x_i \in -W$ for all $i \in I$ with $i \succcurlyeq i_0$. But I is cofinal, so there exists $i \in I$ such that $i \succcurlyeq n_0$ and $i \succcurlyeq i_0$. Then $x_i \in -W$ and therefore $i \in A$. Since $i \succcurlyeq n_0$, it follows that if $n \succcurlyeq n_0$, then $x_n \notin B$, so $x_n \in \neg V$ and therefore $x_n \in -U$. Hence s converges to x. ∎

A quasi-uniform space $(X, \mathcal{U})$ is said to be **uniformly complete** if every uniformly Cauchy net in X converges in the uniform topology to a limit in X. On the other hand, a pre-apartness space X is said to be **totally complete** if every totally Cauchy net in X is apartness convergent to a limit in X.

Proposition 3.5.16 *Every totally complete quasi-uniform apartness space is uniformly complete.*

Proof. Let $(X, \mathcal{U})$ be a totally complete quasi-uniform space, and $(x_n)_{n \in \mathfrak{D}}$ a uniformly Cauchy net in X. Then $(x_n)_{n \in \mathfrak{D}}$ is totally Cauchy, by Proposition 3.5.4, and is therefore $\bowtie_{\mathcal{U}}$-convergent to an element $x \in X$. Given U in $\mathcal{U}$, construct a 3-chain (U, V, W). By Lemmas 3.2.23 and 3.2.21,

$$x \in W[x] \subset - \neg V[x] \subset \neg \neg V[x] \subset U[x].$$

There exists $N \in \mathfrak{D}$ such that $x_n \in - \neg V[x]$, and therefore $x_n \in U[x]$, for all $n \succcurlyeq N$. Since $U \in \mathcal{U}$ is arbitrary, it follows that $(x_n)_{n \in \mathfrak{D}}$ converges to x in the uniform topology on X. ∎

Proposition 3.5.17 *Every uniformly complete, totally bounded uniform space is totally complete.*

Proof. Let the uniform space $(X, \mathcal{U})$ be both uniformly complete and totally bounded. Theorem 3.5.5 shows that every totally Cauchy net in X is uniformly

Cauchy; by the uniform completeness of X, it therefore converges in the uniform topology $\tau_\mathcal{U}$ to a limit in X. Since, by Corollary 3.2.20, $\tau_\mathcal{U}$ coincides with the apartness topology associated with $\bowtie_\mathcal{U}$, the proof is complete. ∎

In view of Corollary 3.2.20, for a uniform space $(X,\mathcal{U})$, when dealing with $\tau_\mathcal{U}$ or $\bowtie_\mathcal{U}$ we may simply and unambiguously speak of *convergence, open, closed, closure*, and the like.

The next two results for uniform spaces are well known; the two after that, for pre-apartness spaces, are not.

Proposition 3.5.18 *Every closed subspace of a uniformly complete uniform space is uniformly complete.*

Proof. Let S be a closed subspace of a uniformly complete uniform space $(X,\mathcal{U})$, and let s be a uniformly Cauchy net in S. By the uniform completeness of X, s converges to a limit x in X. Proposition 2.4.4 ensures that x is in the closure of S and hence in S. ∎

Proposition 3.5.19 *Every uniformly complete subspace of a uniform space is closed.*

Proof. Let S be a uniformly complete subspace of a uniform space $(X,\mathcal{U})$, and let x be in the $\tau_\mathcal{U}$-closure of S. Then

$$\mathcal{D} \equiv \{(\xi,U) : x \in -U, \xi \in S - U\},$$

taken with the reverse inclusion preorder $\succcurlyeq$, is a directed set, and $s : (\xi,U) \rightsquigarrow \xi$ is a net in S that converges to x in X. As we remarked after Corollary 3.5.3, s is uniformly Cauchy in S; whence it converges to a limit $x_\infty \in S$. But X, being a uniform space, is Hausdorff; since, by Corollary 3.2.2, its inequality is tight, Corollary 2.4.8 ensures that $x = x_\infty \in S$. ∎

Proposition 3.5.20 *Every nearly closed subspace of a totally complete pre-apartness space is totally complete.*

Proof. Let S be a nearly closed subspace of a totally complete pre-apartness space X, and let $(x_n)_{n\in\mathfrak{D}}$ be a totally Cauchy net in S. Since X is totally complete, this net converges to a limit $x \in X$. Hence, by Proposition 2.4.4, $x \in \overline{S} = S$. ∎

Proposition 3.5.21 *Every totally complete subspace of a weakly symmetrically separated pre-apartness space with the unique limits property is nearly closed.*

Proof. Let X be a weakly symmetrically separated pre-apartness space with the unique limits property. Let S be a totally complete subspace of X, and x a point of the closure $\overline{S}$ of S in the apartness topology. By Proposition 2.4.4, there exists a net $s \equiv (x_n)_{n\in\mathfrak{D}}$ in S converging to x. It follows from

Proposition 3.5.1 that s is a totally Cauchy net in S; whence, by the total completeness of S, the net s converges to a limit $y \in S$. Since X has the unique limits property, $x = y \in S$. Thus S is closed. ∎

Propositions 3.5.16–3.5.20 suggest that total completeness may be a reasonable lifting of the notion of completeness from uniform spaces to general pre-apartness spaces. This is not the case. For, classically, a uniform space is totally complete if and only if it is compact in the usual sense: every open cover has a finite subcover. Note that this notion of compactness is classically equivalent[1] to the constructive one defined as follows: a uniform space is **(uniformly) compact** if it is both totally bounded and uniformly complete.

In order to prove the classical equivalence between total completeness and compactness, we need more information about filters. With each filter $\mathcal{F}$ on X we associate a net in X as follows. Let

$$\mathfrak{D}_{\mathcal{F}} \equiv \{(x, F) : F \in \mathcal{F} \wedge x \in F\},$$

and define the order relation $\succcurlyeq$ on $\mathfrak{D}_{\mathcal{F}}$ by

$$(x', F') \succcurlyeq (x, F) \Leftrightarrow F' \subset F.$$

If (x, F) and (x', F') are in $\mathfrak{D}_{\mathcal{F}}$, then $F, F' \in \mathcal{F}$, so $F \cap F'$ belongs to $\mathcal{F}$ and therefore contains a point y. Since $(y, F \cap F') \succcurlyeq (x, F)$ and $(y, F \cap F') \succcurlyeq (x', F')$, we see that $\mathfrak{D}_{\mathcal{F}}$ is directed and hence that

$$s_{\mathcal{F}}(x, F) \equiv x$$

defines a net $s_{\mathcal{F}}$ in X.

A filter $\mathcal{F}$ on a topological space (X, τ) is said to **converge to the limit** $x \in X$ if each τ-neighbourhood of x contains some set in $\mathcal{F}$. It is an exercise to show that $\mathcal{F}$ converges to x in X if and only if the net $s_{\mathcal{F}}$ converges to x.

An **ultrafilter** on a set X is a maximal filter on X (relative to set inclusion); in other words, $\mathcal{F}$ is an ultrafilter if and only if

▷ it is a filter on X and

▷ if $\mathcal{F}'$ is a filter on X that includes $\mathcal{F}$, then $\mathcal{F}' = \mathcal{F}$.

For each x in X, the set of all supersets of $\{x\}$ is an ultrafilter. Classically,

▶ a filter $\mathcal{F}$ is an ultrafilter if and only if for each $S \subset X$, either $S \in \mathcal{F}$ or $X \setminus S \in \mathcal{F}$; and

▶ every filter is contained in an ultrafilter.

[1]See [13] (Chapter 1, §9.1, and Chapter 2, §§4.1 and 4.2) or [82] (Chapter 1, Section 7.1, and Chapter 2, Sections 3.6 and 3.7).

For the proof of the former we need **LEM**, and for that of the latter, the axiom of choice (which implies **LEM**). Likewise, **LEM** is essential for the proof of the following classical criterion: a topological space X is compact if and only if every ultrafilter on X converges. Thus, when we deal with ultrafilters, it makes sense for us to use classical, rather than intuitionistic, logic.

We now prove that if $\mathcal{F}$ is an ultrafilter on a topological space (X, τ), then the corresponding net $s_{\mathcal{F}}$ is totally Cauchy. Let A be any cofinal subset of $\mathfrak{D}_{\mathcal{F}}$. Either $s_{\mathcal{F}}(A) \in \mathcal{F}$ or $Y \equiv X \setminus s_{\mathcal{F}}(A) \in \mathcal{F}$. Suppose the latter obtains. Since $\mathcal{F}$ is a filter, there exists $y \in Y$. By the cofinality of A, there exists $(x, F) \in A$ such that $(x, F) \succcurlyeq (y, Y)$; whence

$$s_{\mathcal{F}}(x, F) = x \in F \subset Y = X \setminus s_{\mathcal{F}}(A),$$

which is absurd. Hence, in fact, $s_{\mathcal{F}}(A) \in \mathcal{F}$. If B is any other cofinal set, then $s_{\mathcal{F}}(B) \in \mathcal{F}$ and therefore $s_{\mathcal{F}}(A) \cap s_{\mathcal{F}}(B)$ is inhabited. Thus $s_{\mathcal{F}}(A) \times s_{\mathcal{F}}(B)$ intersects the diagonal of $X \times X$ and therefore intersects each entourage of X. It follows that if A and B are subsets of $\mathfrak{D}_{\mathcal{F}}$ such that $s_{\mathcal{F}}(A) \bowtie s_{\mathcal{F}}(B)$, then at least one of them is not cofinal. This almost trivially implies that the net $s_{\mathcal{F}}$ is totally Cauchy.

It follows from this that if a uniform space $(X, \mathcal{U})$ is totally complete, then every ultrafilter on X converges relative to the uniform topology; whence, as noted earlier, X is (open cover) compact and therefore both complete and totally bounded. However, this classical argument cannot be redeemed constructively: for otherwise it would follow from Propositions 3.5.17 and 3.5.20 that every closed subspace of a uniformly complete, totally bounded metric space is totally bounded. The latter proposition implies the law of excluded middle: consider the subspace

$$\{0\} \cup \{x : x = 1 \wedge (P \vee \neg P)\}$$

of $\{0, 1\}$, where P is any proposition.

To sum up the last few paragraphs: we have shown, *with classical logic and the aid of ultrafilters*, that total completeness and compactness are equivalent conditions on a uniform apartness space. We have also shown, *with intuitionistic logic and a Brouwerian example*, that, even if we take compactness in its constructive sense of total boundedness and uniform completeness, this equivalence cannot be proved constructively.

We now change tack, to produce some results about the extension of continuous functions, beginning with

Theorem 3.5.22 *Let X be a symmetric, weakly symmetrically separated pre-apartness space that has the reverse Kolmogorov property and satisfies* **A4**$_s$. *Let Y be a symmetric, totally complete Hausdorff apartness space that has the Efremovič property, the weak nested neighbourhoods property, and a tight inequality. Let D be a dense subset of X, and let f be a strongly continuous* *mapping of D into Y. Then there exists a strongly continuous mapping $f^{\sharp}$: $X \to Y$ whose restriction to D is f.*

Proof. For each $x \in X$ define

$$\mathfrak{L}_x \equiv \{(\xi, U) : x \in -U \wedge \xi \in D - U\},$$

and furnish $\mathfrak{L}_x$ with the reverse inclusion preorder. For each $n \equiv (\xi, U) \in \mathfrak{L}_x$ let $x_n \equiv \xi$. Then $(x_n)_{n \in \mathfrak{L}_x}$ is a net in X that converges to x; by Proposition 3.5.1, $(x_n)_{n \in \mathfrak{L}_x}$ is therefore a totally Cauchy net. Since f is strongly continuous, $(f(x_n))_{n \in \mathfrak{L}_x}$ is a totally Cauchy net in Y, by Proposition 3.5.14; but Y is totally complete, so this net converges to a limit in Y. Since the inequality on Y is tight and Y is Hausdorff, we see from Corollary 2.4.8 that the limit $f^{\sharp}(x)$ of the net $(f(x_n))_{n \in \mathfrak{L}_x}$ is uniquely determined by x.

We show that if $x \in D$, then $f^{\sharp}(x) = f(x)$. For such x, we can define an (inhabited) cofinal subset

$$\mathfrak{E}_x \equiv \{(x, U) : x \in D - U\},$$

of $\mathfrak{L}_x$. Now the net $(x_n)_{n \in \mathfrak{E}_x}$ converges to x, f is strongly continuous on D, and Y has the property **WNN**; hence the subnet $(f(x_n))_{n \in \mathfrak{E}_x}$ of $(f(x_n))_{n \in \mathfrak{L}_x}$ converges to $f(x)$, by Corollary 2.4.17. But, by Lemma 3.5.8, it also converges to the limit $f^{\sharp}(x)$ of $(f(x_n))_{n \in \mathfrak{L}_x}$. Since the inequality on Y is tight and Y is Hausdorff, Corollary 2.4.8 tells us that $f^{\sharp}(x) = f(x)$.

Now let A and B be subsets of X such that $f^{\sharp}(A) \bowtie f^{\sharp}(B)$. Using **EF** twice and symmetry, construct $E, F \subset Y$ such that $f^{\sharp}(A) \bowtie \neg E$, $E \bowtie \neg F$, and $f^{\sharp}(B) \bowtie F$. For each $x \in A$ we have $f^{\sharp}(x) \in -\neg E$, so there exists $n_0 \in \mathfrak{L}_x$ such that $f(x_n) \in -\neg E$, and therefore $x_n \in f^{-1}(-\neg E \cap f(D))$, for all $n \in \mathfrak{L}_x$ with $n \succcurlyeq n_0$. It follows from Proposition 2.4.4 that

$$x \in \overline{f^{-1}(-\neg E \cap f(D))},$$

where the closure refers to the apartness topology on X. Hence

$$A \subset \overline{f^{-1}(-\neg E \cap f(D))}.$$

Likewise,

$$B \subset \overline{f^{-1}(-F \cap f(D))}.$$

By Proposition 3.1.10, **A4**$_{ss}$ holds in Y. Proposition 3.1.7, together with symmetry, now shows us that $\neg\neg E \bowtie \neg F$; whence $-\neg E \bowtie -F$ and therefore

$$-\neg E \cap f(D) \bowtie -F \cap f(D).$$

Using the strong continuity of f on D, we obtain

$$f^{-1}(-\neg E \cap f(D)) \bowtie f^{-1}(-F \cap f(D)).$$

Since **A4**$_s$ holds in X, it follows from Corollary 3.1.19 that

$$\overline{f^{-1}(-\neg E \cap f(D))} \bowtie \overline{f^{-1}(-F \cap f(D))}.$$

In turn, this implies that $A \bowtie B$. Thus $f^\sharp$ is strongly continuous on X. ∎

Let X be a topological space, and D an inhabited subset of X. We say that D is **strongly dense** in X if for each $x \in X$ the set $D \sim \{x\}$ is dense in X.

Proposition 3.5.23 *Let $(X, \mathcal{U})$ and $(Y, \mathcal{V})$ be uniform spaces, D a strongly dense subset of X, and f a strongly continuous map of X into Y that is uniformly continuous on D. Then f is uniformly continuous on X.*

Proof. Given $x \in X$, define

$$\mathcal{L}'_x \equiv \{(\xi, U) : x \in -U \wedge \xi \in D - U \wedge \xi \neq x\}.$$

Since D is strongly dense, $\mathcal{L}'_x$ is inhabited; taken with the reverse inclusion preorder, it is a directed set. For each $n \equiv (\xi, U) \in \mathcal{L}'_x$ write $x_n \equiv \xi$; then the net $(x_n)_{n \in \mathcal{L}'_x}$ converges to x relative to the apartness on X. Referring to Corollaries 3.2.27 and 2.4.17, we see that the net $(f(x_n))_{n \in \mathcal{L}'_x}$ converges to $f(x)$ in Y. Let V be an entourage of Y, and use Lemma 3.2.17 to construct an open entourage W of Y whose τ_V-closure is a subset of V. Since f is uniformly continuous on D, there exists an entourage U of X such that if $(x, y) \in U \cap (D \times D)$, then $(f(x), f(y)) \in W$. Using Lemma 3.2.15, pick a symmetric open entourage G of X such that $G^3 \subset U$. Given y with $(x, y) \in G$, for each $m \equiv (\eta, V) \in \mathcal{L}'_y$ write $y_m \equiv \eta$. Since uniform apartness spaces are topologically consistent, convergence in the apartness on X is equivalent to convergence relative to its uniform topology. Hence there exist

- $n_0 \in \mathcal{L}'_x$ such that $(x_n, x) \in G$ for all $n \succeq n_0$ in $\mathcal{L}'_x$, and

- $m_0 \in \mathcal{L}'_y$ such that $(y_m, y) \in G$ for all $m \succeq m_0$ in $\mathcal{L}'_y$.

So for all $n \in \mathcal{L}'_x$ with $n \succeq n_0$, and all $m \in \mathcal{L}'_y$ with $m \succeq m_0$, we have $(x_n, y_m) \in G^3 \subset U$. Since also $(x_n, y_m) \in D \times D$, it follows that $(f(x_n), f(y_m)) \in W$. But $(f(x_n))_{n \succeq n_0}$ converges to $f(x)$, and $(f(y_m))_{m \succeq m_0}$ converges to $f(y)$; so each neighbourhood of $(f(x), f(y))$ in the product topology on $Y \times Y$ intersects W. Hence $(f(x), f(y))$ belongs to the closure of W and therefore to V. This proves that f is uniformly continuous on X. ∎

Corollary 3.5.24 *Let X, Y be uniform apartness spaces, where Y is totally complete. Let D be a strongly dense subset of X, and f a uniformly continuous mapping of D into Y. Then f extends to a uniformly continuous mapping of X into Y.*

Proof. Every uniform apartness space is symmetric and has the Efremovič property; whence, by Propositions 3.1.9 and 3.1.10, it has the reverse Kolmogorov property and satisfies **A4**$_s$. It is also weakly symmetrically separated (page 122), has the property **WNN** (page 87), is Hausdorff (page 81), and (by Corollary 3.2.2) has a tight inequality. In view of this and Proposition 3.3.2, we can apply Theorem 3.5.22 to obtain a strongly continuous extension of f to X. Proposition 3.5.23 shows that this extension is uniformly continuous on X. ∎

We end this section with a generalisation of an important result of Bishop about locatedness in metric spaces. We shall deal with locatedness more thoroughly in the next section, but in the meantime we say that a subset S of a pre-apartness space $(X, \bowtie)$ is **weakly located** if

$$\forall_{x \in X} \forall_{U \subset X} (x \in -U \Rightarrow (x \in -S \vee S - U \neq \varnothing)).$$

The geometric interpretation of this property is that for each $x \in X$ and each basic neighbourhood $-U$ of x in the apartness topology, either x is apart from S or else S intersects $-U$; this "either ... or" gives us some feel for the geometric location of x relative to S.

Every located subset S of a metric space (X, ρ) is weakly located. For if $x \in -U$ in X, then there exists $r > 0$ such that $B(x,r) \subset -U$. Either $\rho(x,S) > 0$ and therefore $x \in -S$, or else $\rho(x,S) < r$; in the latter case, there exists $y \in S \cap B(x,r) \subset S - U$.

The following fundamental result about locatedness in metric spaces is known as **Bishop's lemma** ([29], Proposition 3.1.1):

> If S is an inhabited, complete, and located subset of a metric space X, then for each $x \in X$ there exists $y \in S$ such that if $x \neq y$, then $\rho(x,S) > 0$.

We derive an apartness-space counterpart of this very useful result. First, though, we note that if the apartness topology on a pre-apartness space X is **first countable**—that is, if every element of X has a countable base of neighbourhoods in the apartness topology—then for each $x \in X$ there exists a sequence $(U_n)_{n \geqslant 1}$ of subsets of X with the following properties:

▷ $x \in -U_n$ for each n,

▷ for each $U \subset X$ with $x \in -U$, there exists n such that $x \in -U_n \subset -U$;

▷ $-U_1 \supset -U_2 \supset -U_3 \supset \cdots$.

To prove this, consider any $x \in X$ and a countable base $(B_n)_{n \geqslant 1}$ of neighbourhoods of x in the apartness topology. Replacing B_n by $\bigcap_{i=1}^{n} B_i$, we may assume that $B_1 \supset B_2 \supset B_3 \supset \cdots$. Pick a subset U_1 of X such that $x \in -U_1 \subset B_1$. Suppose we have found subsets $U_1, \ldots, U_k$ of X such that

$$x \in -U_k \subset -U_{k-1} \subset \cdots \subset -U_1.$$

Then there exist a positive integer $j > k$ and a subset U_{k+1} of X such that

$$x \in -U_{k+1} \subset B_j \subset -U_k.$$

This completes the inductive construction of a sequence $(U_n)_{n \geqslant 1}$ of subsets of X with the desired properties. Note that this construction uses dependent choice.

We say that a pre-apartness space X is **totally sequentially complete** if every totally Cauchy sequence in X converges to a limit in X. By Corollary 3.5.13, every uniformly complete uniform space is totally sequentially complete.

Proposition 3.5.25 *Let X be a pre-apartness space, and S an inhabited subset of X. If **A5** holds and*

$$\forall_{x \in X} \exists_{y \in S} (x \neq y \Rightarrow x \bowtie S), \tag{3.36}$$

then S is weakly located. Conversely, if X is first countable, weakly symmetrically separated, and Hausdorff, and if S is both totally sequentially complete and weakly located, then (3.36) holds.

Proof. Supposing that (3.36) holds, and given $x \in X$, choose $y \in S$ such that if $x \neq y$, then $x \bowtie S$. If $U \subset X$ and $x \in -U$, then by **A5**, either $x \neq y$ and therefore $x \bowtie S$, or else $y \bowtie U$ and therefore $S - U$ is inhabited. Thus S is weakly located.

Now suppose, conversely, that X is first countable, weakly symmetrically separated, and Hausdorff, and that S is totally sequentially complete and weakly located. Fix an element b of S. Given $x \in X$, choose a sequence $(U_n)_{n \geqslant 1}$ of subsets of X as in the paragraph containing the definition of *first countable*. Using the weak locatedness of S, construct an increasing binary sequence $(\lambda_n)_{n \geqslant 1}$ such that

$$\lambda_n = 0 \Rightarrow S - U_n \neq \varnothing,$$
$$\lambda_n = 1 \Rightarrow x \bowtie S.$$

If $\lambda_1 = 1$, then

$$x \neq b \Rightarrow x \bowtie S.$$

So without loss of generality we may assume that $\lambda_1 = 0$. For each n, if $\lambda_n = 0$, pick $y_n \in S - U_n$; and if $\lambda_n = 1$, set $y_n \equiv y_{n-1}$. We shall show that the sequence $s \equiv (y_n)_{n \geqslant 1}$ obtained in this way is a totally Cauchy sequence. First, however, note that

(*) if $\lambda_n = 0$, then $y_m \in -U_n$ for each $m \geqslant n$.

For, given $m \geqslant n$, we have either $\lambda_m = 0$ and therefore $y_m \in -U_m \subset -U_n$, or else $\lambda_m = 1$. In the latter case, there exists k such that $n < k \leqslant m$, $\lambda_k = 1 - \lambda_{k-1}$, and $y_m = y_{k-1}$; since $\lambda_{k-1} = 0$ and $k > n$, the case just dealt with shows that $y_m \in -U_n$. This completes the proof of (*).

Now let A and B be subsets of $\mathbf{N}^+$ such that $s(A) \bowtie s(B)$. Since X is weakly symmetrically separated, there exists V such that

$$ x \in -V \wedge \neg\,(s(A) - V \neq \varnothing \wedge s(B) - V \neq \varnothing)\,. $$

Pick N such that $-U_N \subset -V$, and suppose that $m, n \geqslant N$ for some $m \in A$ and $n \in B$. If $\lambda_N = 0$, then $y_m \in -U_N$, by (*), and therefore $y_m \in s(A) - U_N$; whence $y_m \in s(A) - V$ and therefore $s(B) - V = \varnothing$. In particular, $y_n \notin -V$, so $y_n \notin -U_N$, which contradicts (*). Thus we must have $\lambda_N = 1$ and therefore $y_m = y_N = y_n$, which is absurd since $s(A) \bowtie s(B)$. This final contradiction shows that s is a totally Cauchy sequence.

Since S is totally sequentially complete, the sequence $(y_n)_{n \geqslant 1}$ converges to a limit $y \in S$. Suppose that $x \neq y$. Since X is Hausdorff, there exist j and W such that $x \in -U_j$ and $y \in -W \subset {\sim}-U_j$; then, by **B4**, $y \bowtie -U_j$. Suppose that $\lambda_j = 0$. If $\lambda_n = 1 - \lambda_{n-1}$ for some $n > j$, then, by (*), $y = y_{n-1} \in -U_j$, a contradiction. Thus $\lambda_n = 0$ for all $n > j$ and therefore for all n. It follows, once again from (*), that $y_k \in -U_n$ for all $k \geqslant n$, so the sequence $(y_n)_{n \geqslant 1}$ converges to x. This is impossible: the apartness topology on X is Hausdorff and therefore, by Proposition 2.4.7, has the strong unique limits property, so $(y_n)_{n \geqslant 1}$ is eventually bounded away from x. Hence the case $\lambda_j = 0$ is ruled out; so $\lambda_j = 1$ and therefore $x \bowtie S$. ∎

Corollary 3.5.26 *Let $(X, \mathcal{U})$ be a first-countable uniform apartness space, and S an inhabited, uniformly complete, weakly located subset of X. Then (3.36) holds.*

Proof. Uniform spaces are both Hausdorff and weakly symmetrically separated. On the other hand, as we observed above, a uniformly complete uniform space is totally sequentially complete. It remains to apply Proposition 3.5.25. ∎

Weak locatedness is one locatedness property that may be valuable in the theory of pre-apartness spaces. We deal with another type of locatedness, applicable to uniform spaces, in the next section.

3.6   Almost Locatedness

We now investigate another type of locatedness for a subset S of a quasi-uniform space $(X, \mathcal{U})$; we say that S is **almost located** if for each $U \in \mathcal{U}$ there exists $V \in \mathcal{U}$ such that

$$ \forall_{x \in X}\,(S \subset \neg V[x] \vee S \cap U[x] \neq \varnothing)\,. \tag{3.37} $$

Note that there is no loss of generality in taking V as a subset of U here: for, since U and V are entourages, so is $U \cap V$, and $\neg V[x] \subset \neg\,(U \cap V)[x]$. Also, we can take V such that (U, V) is a 2-chain of entourages.

Lemma 3.6.1 *A subset S of a quasi-uniform space $(X, \mathcal{U})$ is almost located if and only if for each $U \in \mathcal{U}$, there exists $W \in \mathcal{U}$ such that*

$$\forall_{x \in X} \left(S \subset {\sim} W[x] \vee S \cap U[x] \neq \varnothing \right). \tag{3.38}$$

Proof. It is trivial that the stated condition implies almost locatedness. Conversely, suppose that S is almost located, let $U \in \mathcal{U}$, and construct $V \in \mathcal{U}$ such that (3.37) holds. By Proposition 3.2.6, there exists $W \in U$ such that $X \times X = V \cup {\sim} W$. Then (3.38) follows from (3.37) and the fact that $\neg V[x] \subset {\sim} W[x]$. ∎

A subset S of a metric space (X, ρ) is almost located if and only if for each $\varepsilon > 0$ there exists $\delta \in (0, \varepsilon)$ such that for each $x \in X$, either $S \subset \neg B(x, \delta)$ or $S \cap B(x, \varepsilon)$ is inhabited; in symbols, this condition becomes

$$\forall_{\varepsilon > 0} \exists_{\delta > 0} \forall_{x \in X} \left(\forall_{s \in S} \left(\rho(x, s) \geqslant \delta \right) \vee \exists_{s \in S} \left(\rho(x, s) < \varepsilon \right) \right). \tag{3.39}$$

Proposition 3.6.2 *A located set in a metric space is almost located.*

Proof. Let S be located in the metric space (X, ρ), and let $\varepsilon > 0$. For each $x \in X$, either $\rho(x, S) > \varepsilon/2$ or $\rho(x, S) < \varepsilon$. In the first case, $\rho(x, s) \geqslant \varepsilon/2$ for all $s \in S$; in the second, there exists $s \in S$ with $\rho(x, s) < \varepsilon$. Hence (3.39) holds, and therefore S is almost located. ∎

Proposition 3.6.3 *An almost located subset of a totally bounded uniform space is totally bounded.*

Proof. Let S be an almost located subset of a totally bounded uniform space $(X, \mathcal{U})$, and let $U \in \mathcal{U}$. Construct a 3-chain (U, U_2, U_3) such that U_2 is symmetric and

$$\forall_{x \in X} \left(S \subset \neg U_3[x] \vee S \cap U_2[x] \neq \varnothing \right).$$

Construct points $x_1, \ldots, x_n$ of X such that $X = \bigcup_{i=1}^{n} U_3[x_i]$. Write $\{1, \ldots, n\}$ as a union of sets P, Q such that

$$i \in P \Rightarrow S \cap U_2[x_i] \neq \varnothing,$$
$$i \in Q \Rightarrow S \subset \neg U_3[x_i],$$

and for each $i \in P$ construct $y_i \in S \cap U_2[x_i]$. Consider any $y \in S$. There exists i such that $y \in U_3[x_i]$; so $i \notin Q$ and therefore $i \in P$. Now, $(x_i, y) \in U_3$ and $(y_i, x_i) \in U_2^{-1} = U_2$; whence

$$(y_i, y) \in U_2 \circ U_3 \subset U_2 \circ U_2 \subset U.$$

Thus $S \subset \bigcup_{i \in P} U[y_i]$, from which (since $U \in \mathcal{U}$ is arbitrary) it follows that S is totally bounded. ∎

Corollary 3.6.4 *In a totally bounded metric space, locatedness and almost locatedness coincide.*

Proof. This follows from Propositions 3.6.2 and 3.6.3, together with the fact (see page 15) that totally bounded subsets of a metric space are located. ∎

Here is a converse of Proposition 3.6.3.

Proposition 3.6.5 *A totally bounded subset of a uniform space is almost located.*

Proof. Let S be a totally bounded subset of the uniform space $(X,\mathcal{U})$. Let $U_1 \in \mathcal{U}$, and let (U_1, U_2, U_3) be a 3-chain with U_3 symmetric. Choose $x_1, \ldots, x_n \in S$ such that $S = \bigcup_{i=1}^{n} U_3[x_i]$. Given $x \in X$, we have either

(i) $(x, x_i) \in \neg U_2$ for all i,

or else

(ii) $(x, x_i) \in U_1$, and therefore $x_i \in S \cap U_1[x]$, for some i.

Suppose that (i) applies. If $s \in S$ and $(x, s) \in U_3$, then, choosing i such that $(x_i, s) \in U_3$, we have

$$(x, x_i) \in U_3 \circ U_3^{-1} = U_3^2 \subset U_2,$$

which contradicts (i). It follows that $S \subset \neg U_3[x]$. Putting together the cases (i) and (ii), we have now proved that

$$S \subset \neg U_3[x] \lor S \cap U_1[x] \neq \varnothing.$$

Since $x \in S$ is arbitrary, we conclude that S is almost located. ∎

Proposition 3.6.6 *An almost located subset of a uniform space is weakly located.*

Proof. Let S be an almost located subset of a uniform space $(X, \mathcal{U})$. Let $x \in X$, and let W be a subset of X such that $x \in -W$. There exists a 3-chain (U_1, U_2, U_3) such that $\{x\} \times W \subset \neg U_1$. Since S is almost located, there exists $V \in \mathcal{U}$ such that either $S \subset \neg V[x]$ or $S \cap U_2[x]$ is inhabited. In the first case, $\{x\} \times S \subset \neg V$ and therefore $x \in -S$. On the other hand, if $y \in S \cap U_2[x]$, suppose that $(y, w) \in U_2$ for some $w \in W$; then since $(x, y) \in U_2$, we obtain $(x, w) \in U_2^2 \subset U_1$, which contradicts our original choice of U_1. Hence $\{y\} \times W \in \neg U_2$, and therefore $y \in S - W$. ∎

We shall return to almost located sets shortly; but first we prove some useful technical lemmas, and two results (Corollary 3.6.10 and Theorem 3.6.11) whose metric-space versions have long been important in constructive analysis.

Lemma 3.6.7 *Let* $(X, \mathcal{U})$ *be a uniform space with a countable base of entourages. Then* $(X, \tau_\mathcal{U})$ *is a first-countable topological space. Moreover, there exists a countable base* $(U_n)_{n \geqslant 1}$ *of closed, symmetric entourages of* X *such that* (U_n, U_{n+1}) *is a 2-chain for each* n.

Proof. Let $(V_n)_{n \geqslant 1}$ be a countable base of entourages of X. Then for each $x \in X$, the sequence $(V_n[x])_{n \geqslant 1}$ is a base of neighbourhoods of X in the uniform topology; so $(X, \tau_\mathcal{U})$ is first countable. Using Lemma 3.2.17, choose a closed, symmetric entourage U_1 contained in V_1. To complete the proof, it will suffice to construct, inductively, a sequence $(U_n)_{n \geqslant 1}$ of closed, symmetric entourages of X such that for each n, $U_{n+1} \subset V_n$ and (U_n, U_{n+1}) is a 2-chain. Suppose we have constructed $U_n \in \mathcal{U}$. Since $U_n \cap V_n \in \mathcal{U}$, there exists a symmetric entourage U_{n+1} such that $(U_n \cap V_n, U_{n+1})$ is a 2-chain; in view of Lemma 3.2.17, we can also ensure that U_{n+1} is closed. Then $U_{n+1} \subset V_n$ and (U_n, U_{n+1}) is a 2-chain. This completes the construction of the sequence $(U_n)_{n \geqslant 1}$. ∎

Lemma 3.6.8 *Let* X *be a topological space,* U *a subset of* $X \times X$, *and* x *a point of* X. *Then* $\overline{U[x]} \subset \overline{U}[x]$, *where the closure on the right is relative to the product topology on* $X \times X$.

Proof. Let $y \in \overline{U[x]}$ and let V be any neighbourhood of y in X. Then $V \cap U[x]$ is inhabited, so $\{x\} \times V$ intersects U. It follows that if W is a neighbourhood of x, then $W \times V$ intersects U. Hence every neighbourhood of (x, y) in the product topology intersects U, and therefore $(x, y) \in \overline{U}$. Thus $y \in \overline{U}[x]$. ∎

Lemma 3.6.9 *Let* $(X, \mathcal{U})$ *be a totally bounded uniform space with a countable base* $(U_n)_{n \geqslant 1}$ *of closed, symmetric entourages such that* (U_n, U_{n+1}) *is a 2-chain for each* n. *Let* ξ *be a point of* X, *and* N *a positive integer. Then there exists a closed, totally bounded subset* K *of* X *such that* $U_{N+4}[\xi] \subset K \subset U_N[\xi]$.

Proof. Taking $F_0 \equiv \{\xi\}$, we construct an increasing sequence $(F_k)_{k \geqslant 0}$ of finitely enumerable subsets of X such that for each k,

$$\forall_{x \in F_{k+1}} \exists_{y \in F_k} ((x, y) \in U_{N+k+1}) \tag{3.40}$$

and

$$\forall_{x \in U_{N+4}[\xi]} \exists_{y \in F_k} ((x, y) \in U_{N+k+3}). \tag{3.41}$$

First note that (3.41) holds for $k = 0$. Assume that $F_0, \ldots, F_k$ have been constructed with the applicable properties. Let $\{x_1, \ldots, x_m\}$ be a U_{N+k+4}-approximation to X, and write $\{1, \ldots, m\}$ as a union of subsets A, B such that

$$i \in A \Rightarrow \exists_{y \in F_k} ((x_i, y) \in U_{N+k+1}),$$
$$i \in B \Rightarrow \forall_{y \in F_k} ((x_i, y) \in \neg U_{N+k+2}).$$

Setting

$$F_{k+1} \equiv \{x_i : i \in A\} \cup F_k,$$

we certainly obtain (3.40). Now consider any $x \in U_{N+4}[\xi]$. By our induction hypothesis, there exists $y \in F_k$ with $(x,y) \in U_{N+k+3}$. Choosing i such that $(x_i, x) \in U_{N+k+4}$, we have

$$(x_i, y) \in U_{N+k+4} \circ U_{N+k+3} \subset U_{N+k+3}^2 \subset U_{N+k+2}.$$

Hence i cannot belong to B, so $i \in A$ and therefore $x_i \in F_{k+1}$. Since $(x, x_i) \in U_{N+k+4}^{-1} = U_{N+k+4}$, statement (3.41) holds with k replaced by $k+1$. This completes the inductive construction of the sequence $(F_k)_{k \geqslant 0}$.

Let K be the closure in X of the set

$$S \equiv \bigcup_{n \geqslant 0} F_n.$$

Given $x \in U_{N+4}[\xi]$, let V be a $\tau_\mathcal{U}$-neighbourhood of x in X, and pick k such that $U_{N+k+3}[x] \subset V$. By (3.41), $F_k \cap U_{N+k+3}[x]$, and therefore $F_k \cap V$, is inhabited; whence $S \cap V$ is inhabited. Since x, V are arbitrary, we conclude that $U_{N+4}[\xi] \subset \overline{S} = K$.

Now fix U in $\mathcal{U}$, choose k such that $U_{N+k} \subset U$, and consider any $y \in S$. Pick m such that $y \in F_m$. If $m > k$, then we can use (3.40) to find points

$$y_m = y, y_{m-1} \in F_{m-1}, y_{m-2} \in F_{m-2}, \ldots, y_k \in F_k$$

such that $(y_i, y_{i-1}) \in U_{N+i}$ for $k < i \leqslant m$. Then

$$(y, y_{m-2}) \in U_{N+m} \circ U_{N+m-1} \subset U_{N+m-1}^2 \subset U_{N+m-2},$$

so

$$(y, y_{m-3}) \in U_{N+m-2}^2 \subset U_{N+m-3}.$$

Carrying on in this way, we eventually obtain $(y, y_k) \in U_{N+k}$ and therefore $(y_k, y) \in U_{N+k}^{-1} = U_{N+k}$. Hence

$$y \in \bigcup_{x \in F_k} U_{N+k}[x], \tag{3.42}$$

a statement that also holds when $m \leqslant k$, since then $y \in F_k$. It follows from all this that F_k is a finitely enumerable U_{N+k}-approximation, and therefore U-approximation, to S. Thus S—and therefore (by Proposition 3.3.5) K—is totally bounded. Moreover, taking $k = 0$ in (3.42), we have $S \subset U_N[\xi]$ and therefore, by Lemma 3.6.8, $K \subset \overline{U_N[\xi]} \subset \overline{U_N}[\xi] = U_N[\xi]$. ∎

Corollary 3.6.10 *Let $(X, \mathcal{U})$ be a totally bounded uniform space with a countable base of entourages, and let $U \in \mathcal{U}$. Then there exist closed, totally bounded, U-small sets $K_1, \ldots, K_n$ such that $X = \bigcup_{i=1}^{n} K_i$.*

Proof. By Lemma 3.6.7, there exists a countable base $(U_n)_{n\geqslant 1}$ of closed, symmetric entourages of X such that (U_n, U_{n+1}) is a 2-chain for each n. Pick N such that $U_N^2 \subset U$, and then points $x_1, \ldots, x_n$ of X such that

$$X = \bigcup_{i=1}^{n} U_{N+4}[x_i].$$

Use Lemma 3.6.9 to construct, for each i $(1 \leqslant i \leqslant n)$, a closed, totally bounded set $K_i \subset X$ such that $U_{N+4}[x_i] \subset K_i \subset U_N[x_i]$. Then $X = \bigcup_{i=1}^{n} K_i$. Also,

$$K_i \times K_i \subset U_N^{-1} \circ U_N = U_N^2 \subset U,$$

so K_i is U-small. ∎

Theorem 3.6.11 *Let $(X, \mathcal{U})$ be a totally bounded uniform space with a countable base of entourages, and let f be a strongly continuous mapping of X into $\mathbf{R}$. Then for all but countably many real numbers r, the set*

$$X(f, r) \equiv \{x \in X : f(x) \leqslant r\}$$

is either totally bounded or empty.

Proof. By Theorem 3.3.18, f is uniformly continuous. Using Lemma 3.6.7, construct a countable base $(U_n)_{n\geqslant 1}$ of symmetric entourages of X such that (U_n, U_{n+1}) is a 2-chain for each n. In view of Corollary 3.6.10, for each positive integer k there exist a positive integer n_k and closed, totally bounded, U_k-small sets $X_{k,j}$ $(1 \leqslant j \leqslant n_k)$ such that

$$X = X_{k,1} \cup X_{k,2} \cup \cdots \cup X_{k,n_k}.$$

By Corollary 3.3.8, the infimum of f and, for $k \geqslant 1$ and $1 \leqslant j \leqslant n_k$,

$$c_{k,j} \equiv \inf \{f(x) : x \in X_{k,j}\}$$

are well-defined real numbers. Let $(r_n)_{n\geqslant 1}$ be an enumeration of the set

$$\{\inf f\} \cup \{c_{k,j} : k \geqslant 1 \wedge 1 \leqslant j \leqslant n_k\}.$$

Consider any real number r such that $r \neq r_n$ for each n, and any $U \in \mathcal{U}$. If $r < \inf f$, then $X(f, r) = \varnothing$; so we may assume that $r > \inf f$. Now pick k such that $U_k \subset U$. Write $\{1, \ldots, n_k\}$ as a union of subsets P and Q such that

$$j \in P \Rightarrow c_{k,j} < r,$$
$$j \in Q \Rightarrow c_{k,j} > r.$$

For each $j \in P$ choose $x_{k,j} \in X_{k,j}$ such that $f(x_{k,j}) < r$. Given any $x \in X(f, r)$, pick j such that $x \in X_{k,j}$. Then $c_{k,j} \leqslant f(x) \leqslant r$, so $c_{k,j} < r$ and therefore $j \in P$; whence

$$(x_{k,j}, x) \in X_{k,j} \times X_{k,j} \subset U_k \subset U.$$

Thus

$$\{x_{k,j} : j \in P\}$$

is a finitely enumerable U-approximation to $X(f,r)$. Since U is an arbitrary entourage of X, we conclude that $X(f,r)$ is totally bounded. ∎

For applications of the preceding two results in constructive analysis, see [29].

Lemma 3.6.12 *Let $(X,\mathcal{U})$ be a totally bounded uniform space with a countable base $(U_n)_{n \geq 1}$ of entourages, and let Y be an almost located subset of X. Then there exists a strictly increasing sequence $(n_k)_{k \geq 1}$ of positive integers such that*

$$\forall_{x \in X} \left(Y \subset \neg U_{n_{k+1}}[x] \vee Y \cap U_{n_k}[x] \neq \varnothing \right)$$

for each k.

Proof. Taking $n_1 = 1$, suppose we have found n_k. Since Y is almost located, there exists $V \in \mathcal{U}$ such that

$$\forall_{x \in X} \left(Y \subset \neg V[x] \vee Y \cap U_{n_k}[x] \neq \varnothing \right).$$

To complete the induction, it remains to compute $n_{k+1} > n_k$ such that $U_{n_{k+1}} \subset V$. ∎

Lemma 3.6.13 *Let $(X,\mathcal{U})$ be a totally bounded uniform space with a countable base of entourages, Y an almost located subset of X, and T a totally bounded subset of X. Then there exists a totally bounded set S such that $T \cap Y \subset S \subset Y$.*

Proof. Referring to Lemma 3.6.7, let $(U_n)_{n \geq 1}$ be a countable base of symmetric entourages of X such that (U_n, U_{n+1}) is a 2-chain for each n. Construct a strictly increasing sequence $(n_k)_{k \geq 1}$ of positive integers as in Lemma 3.6.12. For each positive integer k let T_k be a finitely enumerable $U_{n_{k+3}}$-approximation to T. Write T_k as a union of finitely enumerable sets A_k and B_k such that

$$t \in A_k \Rightarrow U_{n_{k+1}}[t] \cap Y \neq \varnothing,$$
$$t \in B_k \Rightarrow Y \subset \neg U_{n_{k+2}}[t].$$

For each t in A_k choose s_t^k in Y such that $(t, s_t^k) \in U_{n_{k+1}}$. Let

$$S_k \equiv \{s_t^k : t \in A_k\},$$

and let S be the closure of $\bigcup_{k \geq 1} S_k$ in the subspace Y. We prove that S is totally bounded. Given an entourage U of X, pick a positive integer m such that $U_{n_m} \subset U$. Consider any positive integer $k \geq m + 2$ and any element s

of S_k. There exists $t \in A_k \subset T_k$ such that $s = s_t^k$ and $(t,s) \in U_{n_{k+1}}$. Also, there exists $y \in T_m$ such that $(y,t) \in U_{n_{m+3}}$. We have

$$(y,s) \in U_{n_{m+3}} \circ U_{n_{k+1}} \subset U_{n_{m+3}}^2 \subset U_{n_{m+2}}.$$

Thus

$$s \in U_{n_{m+2}}[y] \cap S_k \subset U_{n_{m+2}}[y] \cap Y$$

and therefore $y \notin B_m$. Hence $y \in A_m$, $(s_y^m, y) \in U_{n_{m+1}}^{-1} = U_{n_{m+1}}$, and

$$(s_y^m, s) \in U_{n_{m+1}} \circ U_{n_{m+2}} \subset U_{n_{m+1}}^2 \subset U_{n_m} \subset U.$$

Thus $s \in U[s_y^m]$. We now see that

$$\bigcup_{k \geqslant m+2} S_k \subset \bigcup_{x \in S_m} U[x],$$

from which it follows that $\overset{m+1}{\underset{k=1}{\bigcup}} S_k$ is a finitely enumerable U-approximation to $\underset{k \geqslant 1}{\bigcup} S_k$. Thus $\underset{k \geqslant 1}{\bigcup} S_k$, and therefore (by Proposition 3.3.5) its closure S in Y, is totally bounded.

To prove that $T \cap Y \subset S$, let $x \in T \cap Y$, and let U and n_m be as above. Pick $t \in T_m$ such that $(t,x) \in U_{n_{m+3}}$. Then

$$x \in U_{n_{m+3}}[t] \cap Y \subset U_{n_{m+2}}[t] \cap Y$$

and therefore $t \notin B_m$. It follows that $t \in A_m$ and therefore $(t, s_t^m) \in U_{n_{m+1}}$; whence

$$(x, s_t^m) \in U_{n_{m+3}}^{-1} \circ U_{n_{m+1}} = U_{n_{m+3}} \circ U_{n_{m+1}} \subset U_{n_{m+1}}^2 \subset U_{n_m}$$

and so

$$s_t^m \in U_{n_m}[x] \cap Y \subset U[x] \cap Y.$$

Thus every neighbourhood of x in Y intersects S, and therefore (S being closed in Y) x is in S. ∎

We say that an inhabited subset Y of a uniform space $(X, \mathcal{U})$ is **weak-locally totally bounded** if there exists an entourage U of X such that for each $x \in X$, there exists a totally bounded subset of Y that contains $Y \cap U[x]$. Referring to remarks on page 15, we readily see that a locally totally bounded metric space is weak-locally totally bounded. The following result is a uniform-space analogue of a useful theorem about locally totally bounded metric spaces ([29], Proposition 2.2.18).

Theorem 3.6.14 *The following hold for an inhabited subset Y of a uniform space X:*

(i) *If Y is weak-locally totally bounded, then it is almost located.*

(ii) *If X is weak-locally totally bounded and has a countable base of entourages, and Y is almost located, then Y is weak-locally totally bounded.*

Proof. First take Y to be weak-locally totally bounded, and construct $U_0 \in \mathcal{U}$ such that for each x in X, there exists a totally bounded subset of Y that contains $Y \cap U_0[x]$. Let U be any entourage of X, let $U_1 \equiv U \cap U_0$, and construct a 3-chain (U_1, U_2, U_3) with U_3 symmetric. By our choice of U_0, for each x in X there exist $y_1, \ldots, y_n \in Y$ such that

$$Y \cap U_0[x] \subset \bigcup_{i=1}^{n} U_3[y_i].$$

Either $(x, y_i) \in U_1$ for some i or else $(x, y_i) \notin U_2$ for all i. In the first case, $Y \cap U_1[x]$ is inhabited, as therefore is $Y \cap U[x]$. In the second case, if $y \in Y \cap U_3[x]$, then since $y \in Y \cap U_0[x]$, there exists i such that $y \in U_3[y_i]$ and therefore $(y, y_i) \in U_3^{-1} = U_3$; but $(x, y) \in U_3$, so $(x, y_i) \in U_3^2 \subset U_2$, a contradiction. We conclude that in this case, $Y \cap U_3[x] = \varnothing$ and hence $Y \subset \neg U_3[x]$. This completes the proof of part (i) of our theorem.

To prove part (ii), suppose that X is weak-locally totally bounded and has a countable base of entourages, and that Y is almost located. Choose $U_0 \in \mathcal{U}$ such that for each $x \in X$, there exists a totally bounded subset of X that contains $U_0[x]$. Given y in Y, construct a totally bounded set $T \subset X$ such that $U_0[y] \subset T$. By Lemma 3.6.13, there exists a totally bounded set $S \subset Y$ such that $T \cap Y \subset S$. Then $U_0[y] \cap Y \subset T \cap Y \subset S$. Since $y \in Y$ is arbitrary, we conclude that Y is weak-locally totally bounded. ∎

3.7 Product Apartness Spaces

The definition of the product of two set-set apartness spaces is much more complicated than that for point-set apartness spaces. If $X \equiv X_1 \times X_2$, where X_1 and X_2 are apartness spaces, then we define the **product apartness** on X as follows. Two subsets A, B of X are **apart**, in which case we write $A \bowtie B$, if there exist finitely many subsets A_i $(1 \leqslant i \leqslant m)$ and B_j $(1 \leqslant j \leqslant n)$ of X such that

▶ $A \subset A_1 \cup \cdots \cup A_m$,

▶ $B \subset B_1 \cup \cdots \cup B_n$, and

▶ for all i, j either $\mathrm{pr}_1 A_i \bowtie \mathrm{pr}_1 B_j$ or $\mathrm{pr}_2 A_i \bowtie \mathrm{pr}_2 B_j$.

We shall prove shortly that this prescription does indeed give us an apartness on X. We then call X, taken with its standard inequality and the apartness just defined, the **product of the apartness spaces** X_1 and X_2. Note that, replacing each A_i by $A \cap A_i$ and each B_j by $B \cap B_j$, we may replace the symbol $\subset$ by $=$ in the foregoing definition.

Proposition 3.7.1 *Let* X_1, X_2 *be apartness spaces, and let* **apart** *denote the product point-set apartness on* $X \equiv X_1 \times X_2$. *Then* $\{\mathbf{x}\} \bowtie S$ *in the product set-set apartness space* X *if and only if* $x \bowtie_{\tau_u} S$.

Proof. Assume first that $\{\mathbf{x}\} \bowtie S$. Then there exist subsets A and B_j $(1 \leqslant j \leqslant n)$ of X such that $\mathbf{x} \in A$, $S \subset B_1 \cup \cdots \cup B_n$, and for all j, either $x_1 \bowtie \mathrm{pr}_1 B_j$ or $x_2 \bowtie \mathrm{pr}_2 B_j$. We may assume without loss of generality that $x_1 \bowtie \mathrm{pr}_1 B_j$ for $1 \leqslant j \leqslant \nu$, and $x_2 \bowtie \mathrm{pr}_2 B_j$ for $\nu + 1 \leqslant j \leqslant n$. Write

$$U \equiv \mathrm{pr}_1 \left(B_1 \cup \cdots \cup B_\nu \right),$$
$$V \equiv \mathrm{pr}_2 \left(B_{\nu+1} \cup \cdots \cup B_n \right).$$

Then, by **B3**, $x_1 \in -U$ and $x_2 \in -V$. Given $x_1' \in -U, x_2' \in -V$, and $\mathbf{s} \equiv (s_1, s_2) \in S$, choose j such that $\mathbf{s} \in B_j$. If $1 \leqslant j \leqslant \nu$, then $x_1' \neq s_1$; if $\nu + 1 \leqslant j \leqslant n$, then $x_2' \neq s_2$. Hence $(x_1', x_2') \neq (s_1, s_2)$. We now see that

$$\mathbf{x} \in -U \times -V \subset \sim S$$

—in other words, $x \bowtie_{\tau_u} S$.

Now assume, conversely, that $x \bowtie_{\tau_u} S$; so there exist $U_i \subset X_i$ with

$$(x_1, x_2) \in -U_1 \times -U_2 \subset \sim S.$$

Since, being a set-set *apartness* space, X_i is locally decomposable, we can find $V_i \subset X_i$ such that $x_i \in -V_i$ and $X = -U_i \cup V_i$. Set

$$A \equiv \{\mathbf{x}\}, \quad B_1 \equiv V_1 \times X_2, \quad B_2 \equiv X_1 \times V_2.$$

For any $\mathbf{s} \in S$ we have $\mathbf{s} \notin -U_1 \times -U_2$; whence either $s_1 \in V_1$ or $s_2 \in V_2$. It follows that $S \subset B_1 \cup B_2$. On the other hand, since $x_i \in -V_i$, we have

$$\mathrm{pr}_i(A) = \{x_i\} \bowtie V_i = \mathrm{pr}_i(B_i).$$

Hence $\{\mathbf{x}\} \bowtie S$. ∎

Having shown that our definition of *product set-set apartness* is compatible with the earlier definition of *product point-set apartness*, we still need to verify that, in the light of axioms **B1–B5**, we have actually defined a set-set apartness on the Cartesian product of the apartness spaces X_1 and X_2.

Proposition 3.7.2 *Let* $X \equiv X_1 \times X_2$ *be the product of two apartness spaces. Then the relation* $\bowtie$, *defined as above for* X, *satisfies axioms* **B1–B5**.

Proof. To verify **B1**, take $A \equiv X$ and $B \equiv \varnothing \subset X$; then, applying **B1** in the space X_1, we obtain $\mathrm{pr}_1(A) = X_1 \bowtie \varnothing = \mathrm{pr}_1(B)$. For **B2** we refer to Proposition 3.7.1. To verify **B3**, consider subsets R,S,T of X. It is easy to prove that if $R \bowtie (S \cup T)$, then $R \bowtie S$ and $R \bowtie T$. Suppose, conversely, that $R \bowtie S$ and $R \bowtie T$. There exist subsets $R_1,\ldots,R_p,S_1,\ldots,S_m$ of X such that $R \subset R_1 \cup \cdots \cup R_p$, $S \subset S_1 \cup \cdots \cup S_m$, and for each pair (i,j) there exists $k \in \{1,2\}$ with $\mathrm{pr}_k(R_i) \bowtie \mathrm{pr}_k(S_j)$ in X_k. Likewise, there exist subsets $R'_1,\ldots,R'_q,T_1,\ldots,T_n$ of X such that $R \subset R'_1 \cup \cdots \cup R'_q$, $T \subset T_1 \cup \cdots \cup T_n$, and for each pair (i,j) there exists $k \in \{1,2\}$ with $\mathrm{pr}_k(R'_i) \bowtie \mathrm{pr}_k(T_j)$ in X_k. Let

$$P_{i,j} \equiv R_i \cap R'_j \quad (1 \leqslant i \leqslant p,\ 1 \leqslant j \leqslant q).$$

Then

$$R \subset \bigcup_{\substack{1 \leqslant i \leqslant p \\ 1 \leqslant j \leqslant q}} P_{i,j}.$$

If $1 \leqslant i \leqslant p$, $1 \leqslant j \leqslant q$, and $1 \leqslant l \leqslant m$, then, choosing k such that $\mathrm{pr}_k(R_i) \bowtie \mathrm{pr}_k(S_l)$, we see that

$$\mathrm{pr}_k(R_i) \cap \mathrm{pr}_k(R'_j) \bowtie \mathrm{pr}_k(S_l),$$

in X_k; whence $\mathrm{pr}_k(P_{i,j}) \bowtie \mathrm{pr}_k(S_l)$ in X_k. Likewise, if $1 \leqslant l \leqslant n$, we can find k such that $\mathrm{pr}_k(P_{i,j}) \bowtie \mathrm{pr}_k(T_l)$. Since

$$S \cup T \subset \bigcup_{1 \leqslant i \leqslant m} S_i \cup \bigcup_{1 \leqslant j \leqslant n} T_j,$$

it follows from the definition of the product apartness that $R \bowtie S \cup T$. Similar arguments show that $R \cup S \bowtie T$ if and only if $R \bowtie T$ and $S \bowtie T$. Putting all this together, we readily complete the verification of **B3**.

Property **B4** holds in X because of Proposition 3.7.1. Referring to that proposition and to Proposition 2.5.6, we see, finally, that **B5** holds in X. ∎

Corollary 3.7.3 *Let* $X \equiv X_1 \times X_2$ *be the product of two apartness spaces. Then the projection maps* $\mathrm{pr}_k : X \to X_k$ *are strongly continuous. Moreover, if* $\bowtie'$ *is any pre-apartness on* X *relative to which the projection maps are strongly continuous, then*

$$\forall_{A,B \subset X} (A \bowtie B \Rightarrow A \bowtie' B).$$

Proof. Let $S \bowtie T$ in X_1, so that $\mathrm{pr}_1(\mathrm{pr}_1^{-1}(S)) \bowtie \mathrm{pr}_1(\mathrm{pr}_1^{-1}(T))$ in X_1. It follows from the definition of the apartness on the product space that $\mathrm{pr}_1^{-1}(S) \bowtie \mathrm{pr}_1^{-1}(T)$ in X. Thus pr_1, and likewise pr_2, is strongly continuous.

Now consider any pre-apartness $\bowtie'$ on X with respect to which the projections $\mathrm{pr}_k : X \to X_k$ are strongly continuous. Given $A,B \subset X$ with $A \bowtie B$, pick the sets A_i,B_j as in the definition of *product apartness*. For each pair (i,j) there exists $k \in \{1,2\}$ such that $\mathrm{pr}_k(A_i) \bowtie \mathrm{pr}_k(B_j)$; hence, by our hypotheses,

$\mathrm{pr}_k^{-1}(\mathrm{pr}_k(A_i)) \bowtie' \mathrm{pr}_k^{-1}(\mathrm{pr}_k(B_j))$, and therefore $A_i \bowtie' B_j$. It follows from this and **B3** that $S \bowtie' T$. ■

Recall, from page 94, the definition of the product of two uniform spaces.

Corollary 3.7.4 *Let $X \equiv X_1 \times X_2$ be the product of two uniform apartness spaces, and A,B subsets of X that are apart relative to the product apartness. Then A,B are apart relative to the product uniform structure on X.*

Proof. In view of Corollary 3.7.3, it is enough to note, as on page 94, that the projections on X are strongly continuous relative to the apartness induced by the product uniform structure. ■

A special case of the preceding corollary occurs when X_1 and X_2 are metric apartness spaces: if two subsets of $X_1 \times X_2$ are apart relative to the product apartness, then they are apart relative to the product metric. The converse is *not* true: consider the case $X_1 = X_2 \equiv \mathbf{Z}$, with the standard metric apartness, and let

$$A \equiv \{(z_1, z_2) : z_1, z_2 \in \mathbf{Z}, \ z_1 \neq z_2\},$$
$$B \equiv \{(z_1, z_2) : z_1, z_2 \in \mathbf{Z}, \ z_1 = z_2\}.$$

Then for all $(z_1, z_2) \in A$ and $(z, z) \in B$,

$$\rho((z_1, z_2), (z, z)) = \max\{|z_1 - z|, |z_2 - z|\} \geqslant 1.$$

Thus A and B are apart relative to the product metric on $\mathbf{Z} \times \mathbf{Z}$. Now suppose that they are apart relative to the product apartness $\bowtie$ on $\mathbf{Z} \times \mathbf{Z}$. Then there exist subsets $A_1, \ldots, A_m$ and $B_1, \ldots, B_n$ of $\mathbf{Z} \times \mathbf{Z}$ such that

- $A \subset A_1 \cup \cdots \cup A_m$,

- $B \subset B_1 \cup \cdots \cup B_n$, and

- for each pair (i, j), there exists $k \in \{1, 2\}$ such that $\mathrm{pr}_k(A_i) \bowtie \mathrm{pr}_j(B_j)$ in $\mathbf{Z}$.

For some $j \leqslant n$ there exist distinct integers z_1, z_2 such that (z_1, z_1) and (z_2, z_2) belong to B_j. Then $(z_1, z_2) \in A$, so there exists i with $(z_1, z_2) \in A_i$. Pick $k \in \{1, 2\}$ such that $\mathrm{pr}_k(A_i) \bowtie \mathrm{pr}_k(B_j)$. Then

$$z_k = \mathrm{pr}_k(z_1, z_2) \neq \mathrm{pr}_k(z_k, z_k) = z_k,$$

which is absurd.

Next we see that the product set-set apartness structure has the characteristic property of a categorical product (cf. [44], page 47).

Proposition 3.7.5 *Let* $X \equiv X_1 \times X_2$ *be the product of two apartness spaces, and* f *a mapping of an apartness space* Y *into* X. *Then* f *is strongly continuous if and only if* $\mathrm{pr}_i \circ f$ *is strongly continuous for each* i.

Proof. Assume that $\mathrm{pr}_i \circ f$ is strongly continuous for each i. Let S,T be subsets of Y such that $f(S) \bowtie f(T)$, and choose finitely many subsets A_i ($1 \leqslant i \leqslant m$) and B_j ($1 \leqslant j \leqslant n$) of X such that

> $f(S) \subset A_1 \cup \cdots \cup A_m$,

> $f(T) \subset B_1 \cup \cdots \cup B_n$, and

> for all i,j either $\mathrm{pr}_1 A_i \bowtie \mathrm{pr}_1 B_j$ or $\mathrm{pr}_2 A_i \bowtie \mathrm{pr}_2 B_j$.

For all i,j write $S_i \equiv f^{-1}(A_i)$ and $T_j \equiv f^{-1}(B_j)$. Then $S \subset S_1 \cup \cdots \cup S_m$ and $T \subset T_1 \cup \cdots \cup T_n$. Also, for all i,j we have either $\mathrm{pr}_1 \circ f(S_i) \bowtie \mathrm{pr}_1 \circ f(T_j)$ or $\mathrm{pr}_2 \circ f(S_i) \bowtie \mathrm{pr}_2 \circ f(T_j)$; whence $S_i \bowtie T_j$, by our strong continuity hypothesis. Applying **B3**, we obtain

$$\bigcup_{i=1}^{m} S_i \bowtie \bigcup_{j=1}^{n} T_j$$

and therefore $S \bowtie T$. Hence f is strongly continuous.

The converse readily follows from the continuity of the projection maps (Corollary 3.7.3) and the preservation of strong continuity under the composition of mappings. ∎

Proposition 3.7.6 *For* $k \in \{1,2\}$, *let* $f_k : X_k \to Y_k$ *be a strongly continuous mapping between apartness spaces. Then*

$$(f_1, f_2) : (x_1, x_2) \rightsquigarrow (f_1(x_1), f_2(x_2))$$

is a strongly continuous mapping between the product apartness spaces $X_1 \times X_2$ *and* $Y_1 \times Y_2$.

Proof. We first observe that, for $k \in \{1,2\}$,

$$\mathrm{pr}_k \circ (f_1, f_2) = f_k = f_k \circ \mathrm{pr}_k.$$

By Corollary 3.7.3, $\mathrm{pr}_k : X_1 \times X_2 \to X_k$ is strongly continuous. Hence the composition $f_k \circ \mathrm{pr}_k$ is strongly continuous, and so $\mathrm{pr}_k \circ (f_1, f_2)$ is strongly continuous. It now follows from Proposition 3.7.5 that (f_1, f_2) is strongly continuous. ∎

Proposition 3.7.7 *A net in a product apartness space is totally Cauchy if and only if its projections onto the factors are totally Cauchy.*

Proof. Let $X \equiv X_1 \times X$ be the product of two apartness spaces, and $s \equiv (\mathbf{x}_n)_{n \in \mathcal{D}}$ a net in X. If this net is totally Cauchy, then Corollary 3.7.3 and

Proposition 3.5.14 together show that $(\mathrm{pr}_k(\mathbf{x}_n))_{n\in\mathfrak{D}}$ is a totally Cauchy net in X_k.

Conversely, suppose that each of the nets $(\mathrm{pr}_k(\mathbf{x}_n))_{n\in\mathfrak{D}}$ is totally Cauchy. Let P,Q be subsets of $\mathfrak{D}$ such that $s(P) \bowtie s(Q)$ in X. There exist subsets $A_1,\ldots,A_p$ and $B_1,\ldots,B_q$ of X such that

- $s(P) \subset A_1 \cup A_2 \cup \cdots \cup A_p$,

- $s(Q) \subset B_1 \cup B_2 \cup \cdots \cup B_q$, and

- for each applicable pair (i,j), there exists $k \in \{1,2\}$ such that $\mathrm{pr}_k(A_i) \bowtie \mathrm{pr}_k(B_j)$ in X_k.

For such i,j, and k,

$$\mathrm{pr}_k \circ s \left(s^{-1}(A_i)\right) \bowtie \mathrm{pr}_k \circ s \left(s^{-1}(B_j)\right),$$

so, since $(\mathrm{pr}_k(\mathbf{x}_n))_{n\in\mathfrak{D}}$ is a totally Cauchy net in X_k, there exists $n_{ij} \in \mathfrak{D}$ such that if $m \in s^{-1}(A_i)$ for some $m \succcurlyeq n_{ij}$, then $n \not\succcurlyeq n_{ij}$ for all $n \in s^{-1}(B_j)$. Construct $N \in \mathfrak{D}$ such that

$$N \succcurlyeq n_{ij} \ \ (1 \leqslant i \leqslant p, \ 1 \leqslant j \leqslant q).$$

Suppose there exists $m \in P$ such that $m \succcurlyeq N$. Choose i such that $s(m) \in A_i$. For each $n \in Q$ choose j such that $n \in s^{-1}(B_j)$; since $m \succcurlyeq n_{ij}$ we must have $n \not\succcurlyeq n_{ij}$ and therefore $n \not\succcurlyeq N$. Thus s is a totally Cauchy net in X. ∎

Proposition 3.7.8 *A product of two apartness spaces is totally complete if and only if each of its factors is totally complete.*

Proof. Let $X \equiv X_1 \times X_2$ be the product of two apartness spaces. Suppose first that each X_k is totally complete, and consider a totally Cauchy net $s \equiv (\mathbf{x}_n)_{n\in\mathfrak{D}}$ in X. By Proposition 3.7.7, for $k \in \{1,2\}$, $\mathrm{pr}_k \circ s$ is a totally Cauchy net in X_k and so converges to a limit ξ_k in X_k. Let $\xi \equiv (\xi_1,\xi_2) \in -U$ in X, and choose $U_k \subset X_k$ such that $\xi \in -U_1 \times -U_2 \subset -U$. There exists $n_k \in \mathfrak{D}$ such that $\mathrm{pr}_k(\mathbf{x}_n) \in -U_k$ for all $n \succcurlyeq n_k$. Pick $N \in \mathfrak{D}$ such that $N \succcurlyeq n_1$ and $N \succcurlyeq n_2$. For all $n \succcurlyeq N$ we have $\mathbf{x}_n \in -U_1 \times -U_2$, so $\mathbf{x}_n \in -U$. Thus s converges to ξ in X.

Now suppose, conversely, that X is totally complete. This time, consider a totally Cauchy net $s_1 \equiv (x_n)_{n\in\mathfrak{D}}$ in X_1. Fixing $\xi_2 \in X_2$, we see from Proposition 3.7.7 that $((x_n,\xi_2))_{n\in\mathfrak{D}}$ is a totally Cauchy net in X; it therefore converges to a limit ξ in X. Let $\xi_1 \in -U_1$ in X_1. Then (see Lemma 2.5.2) $\xi \in -U_1 \times X_2 = -(U_1 \times X_2)$, so there exists N such that $(x_n,\xi_2) \in -U_1 \times X_2$, and therefore $x_n \in -U_1$, for all $n \succcurlyeq N$. Hence ξ_1 is a limit of the net s in X_1. Thus X_1, and similarly X_2, is totally complete. ∎

The next two little lemmas will help us investigate the transfer of the Efremovič property from apartness spaces to their product, and vice versa.

Lemma 3.7.9 *Let $X \equiv X_1 \times X_2$ be the product of two apartness spaces, and let $y \in X_2$. Then the mapping $f : x \rightsquigarrow (x, y)$ is strongly continuous on X_1.*

Proof. Since $\mathrm{pr}_1 \circ f$ (the identity function on X_1) and $\mathrm{pr}_2 \circ f$ (a constant function) are strongly continuous, Proposition 3.7.5 shows that f is strongly continuous. ∎

Lemma 3.7.10 *Let $X \equiv X_1 \times X_2$ be the product of two apartness spaces, let S and T be subsets of X_1, and let $y \in X_2$. Then $S \bowtie T$ in X_1 if and only if $S \times \{y\} \bowtie T \times \{y\}$ in X.*

Proof. If $S \bowtie T$ in X_1, then it follows more or less immediately from the definition of the product apartness that $S \times \{y\} \bowtie T \times \{y\}$ in X. The converse follows from the strong continuity of the mapping $x \rightsquigarrow (x, y)$ on X_1 (Lemma 3.7.9). ∎

Proposition 3.7.11 *If the product of two apartness spaces has the Efremovič property, then so does each of its factors.*

Proof. Let X_1, X_2 be apartness spaces, and suppose that their product X has the Efremovič property. Pick $y \in X_2$, and let S, T be subsets of X_1 such that $S \bowtie T$. Then $S \times \{y\} \bowtie T \times \{y\}$ in X, by Lemma 3.7.10. Applying **EF** in X, construct $E \subset X$ such that $S \times \{y\} \bowtie \neg E$ and $E \bowtie T \times \{y\}$. Let $f : X_1 \to X$ be defined by $f(x) \equiv (x, y)$. Since, by Lemma 3.7.9, f is strongly continuous, Lemma 3.7.10 shows that $S \bowtie f^{-1}(\neg E)$ and $f^{-1}(E) \bowtie T$ in X_1. But $f^{-1}(\neg E) = \neg f^{-1}(E)$, so $S \bowtie \neg f^{-1}(E)$ in X_1. Thus X_1, and similarly X_2, satisfies **EF**. ∎

Here is a partial converse to the preceding proposition.

Proposition 3.7.12 *Let X_1, X_2 be apartness spaces that have the Efremovič property. Then the following conditions are equivalent on $X \equiv X_1 \times X_2$:*

(i) *X satisfies **EF**.*

(ii) *X satisfies **A4**$_{ss}$.*

(iii) *X satisfies **A4**$_s$.*

(iv) *$\forall_{A,S,T \subset X} (A \bowtie (\neg S \cup \neg T) \Rightarrow A \bowtie \neg(S \cap T))$.*

Proof. We already know from Propositions 3.1.10 and 3.1.8 that (i) implies (ii) and that (iii) implies (iv); moreover, it is clear that (ii) implies (iii). To prove that (iv) implies (i), suppose that (iv) holds, and let $S \bowtie T$ in X. Construct subsets A_i ($1 \leqslant i \leqslant m$) and B_j ($1 \leqslant j \leqslant n$) of X such that $S \subset A_1 \cup \cdots \cup A_m$, $T \subset B_1 \cup \cdots \cup B_n$, and for each applicable pair (i, j), there exists $k(i, j)$ such that $\mathrm{pr}_{k(i,j)}(A_i) \bowtie \mathrm{pr}_{k(i,j)}(B_j)$. Applying **EF** in the space $X_{k(i,j)}$, we obtain $F_{ij} \subset X_{k(i,j)}$ such that $\text{pr}_{k(i,j)}(A_i) \bowtie \neg F_{ij}$ and $F_{ij} \bowtie \text{pr}_{k(i,j)}(B_j)$. Writing

$$E_i \equiv \bigcap_{j=1}^{n} \text{pr}_{k(i,j)}^{-1}(F_{ij}) \ \text{ and } \ E \equiv \bigcup_{i=1}^{m} E_i,$$

we have $\text{pr}_{k(i,j)}(E_i) \subset F_{ij}$ and therefore $\text{pr}_{k(i,j)}(E_i) \bowtie \text{pr}_{k(i,j)}(B_j)$. It follows from the definition of *product apartness* that $E \bowtie T$. On the other hand, for all i and j, since $\text{pr}_{k(i,j)}(A_i) \bowtie \neg F_{ij}$, it follows from the strong continuity of $\text{pr}_{k(i,j)}$ that $A_i \bowtie \text{pr}_{k(i,j)}^{-1}(\neg F_{ij})$ in X. Hence, by **B3** in X,

$$A_i \bowtie \bigcup_{j=1}^{n} \neg \text{pr}_{k(i,j)}^{-1}(F_{ij}).$$

Applying (iv), we now obtain

$$A_i \bowtie \neg \bigcap_{j=1}^{n} \text{pr}_{k(i,j)}^{-1}(F_{ij}) = \neg E_i \supset \neg E$$

for each i; whence, again by **B3**, $S \bowtie \neg E$. This completes the proof that (iv) implies (i). ∎

We call a pre-apartness $\bowtie$ on a set X **stable** if

$$\forall_{S,T \subset X} (S \bowtie T \Rightarrow \neg\neg S \bowtie \neg\neg T),$$

in which case, by Lemma 3.1.1,

$$\forall_{S,T \subset X} (S \bowtie T \Leftrightarrow S \bowtie \neg\neg T).$$

Lemma 3.7.13 *If a symmetric pre-apartness space satisfies* **A4**$_{ss}$*, then the pre-apartness is stable.*

Proof. This is a simple application of Proposition 3.1.7 and the symmetry of the pre-apartness. ∎

Proposition 3.7.14 *The product of two symmetric apartness spaces with the Efremovič property has that property if and only if the product apartness is stable.*

Proof. Let X_1, X_2 be symmetric apartness spaces, and X their product. Suppose that X has the Efremovič property. By Proposition 3.1.10, X satisfies **A4**$_{ss}$; since the product apartness is symmetric, it follows from Lemma 3.7.13 that it is stable.

Assuming, conversely, that the product apartness $\bowtie$ is stable, let $S \bowtie T$ in X. Construct subsets A_i $(1 \leqslant i \leqslant m)$ and B_j $(1 \leqslant j \leqslant n)$ of X such that $S \subset A_1 \cup \cdots \cup A_m$, $T \subset B_1 \cup \cdots \cup B_n$, and for each applicable pair (i,j), there exists $k(i,j)$ such that $\text{pr}_{k(i,j)}(A_i) \bowtie \text{pr}_{k(i,j)}(B_j)$. Applying **EF** in the space $X_{k(i,j)}$, construct $F_{ij} \subset X_{k(i,j)}$ such that $\mathrm{pr}_{k(i,j)}(A_i) \bowtie \neg F_{ij}$ and $F_{ij} \bowtie \mathrm{pr}_{k(i,j)}(B_j)$. Now write

$$E_i \equiv \bigcap_{j=1}^{n} \neg \neg \mathrm{pr}_{k(i,j)}^{-1}(F_{ij}),$$

$$E \equiv \bigcup_{i=1}^{m} E_i.$$

Then for all i and j we have

$$E_i \subset \neg \neg \mathrm{pr}_{k(i,j)}^{-1}(F_{ij}) = \mathrm{pr}_{k(i,j)}^{-1}(\neg \neg F_{ij})$$

and therefore $\mathrm{pr}_{k(i,j)}(E_i) \subset \neg \neg F_{ij}$. Since (by Proposition 3.1.10) $X_{k(i,j)}$ satisfies **A4**$_{ss}$, its apartness is stable, by Lemma 3.7.13; so $\neg \neg F_{ij} \bowtie \mathrm{pr}_{k(i,j)}(B_j)$ and therefore $\mathrm{pr}_{k(i,j)}(E_i) \bowtie \mathrm{pr}_{k(i,j)}(B_j)$. It follows from the definition of *product apartness* that $E \bowtie T$. On the other hand, for all i and j, since $\mathrm{pr}_{k(i,j)}(A_i) \bowtie \neg F_{ij}$, it follows from the strong continuity of $\mathrm{pr}_{k(i,j)}$ that $A_i \bowtie \mathrm{pr}_{k(i,j)}^{-1}(\neg F_{ij})$; whence, by **B3**,

$$A_i \bowtie \bigcup_{j=1}^{n} \mathrm{pr}_{k(i,j)}^{-1}(\neg F_{ij}).$$

The stability of $\bowtie$ now yields

$$A_i \bowtie \neg \neg \bigcup_{j=1}^{n} \mathrm{pr}_{k(i,j)}^{-1}(\neg F_{ij}).$$

But

$$\neg \neg \bigcup_{j=1}^{n} \mathrm{pr}_{k(i,j)}^{-1}(\neg F_{ij}) = \neg \bigcap_{j=1}^{n} \neg \mathrm{pr}_{k(i,j)}^{-1}(\neg F_{ij}) = \neg E_i,$$

so $A_i \bowtie \neg E_i \supset \neg E$. Applying **B3** once again, we obtain $S \bowtie \neg E$, which completes the proof that X satisfies **EF**. ∎

3.8   Proximal Connectedness

For a topological space (X, τ) there is at least one constructive concept of connectedness: namely, that if S, T are inhabited open sets whose union is X, then $S \cap T$ is inhabited. The analysis of this concept for metric spaces readily generalises to the setting of a topological space, but we shall not discuss it, or other metric-based notions of connectedness, further. Instead, we concentrate on a type of connectedness that applies to pre-apartness spaces.

We say that a pre-apartness space $(X, \bowtie)$ is **proximally connected** if for each $S \subset X$ with $S \bowtie \neg S$ we have $S \times \neg S = \varnothing$. Note that if $S \times \neg S = \varnothing$, it does not follow constructively that either $S = \varnothing$ or $\neg S = \varnothing$: consider, in the space $\{0\}$, the subset

$$S \equiv \{x : x = 0 \wedge P\},$$

where P is any mathematical statement.

A first example of a proximally connected set is given by the following result:

Proposition 3.8.1 *Every convex subset of a normed space is proximally connected.*

Proof. Let C be a convex subset of a normed space X, and let S be a subset of C such that $S \bowtie (C \cap \neg S)$. Since X is a (metric) uniform space, we see from Proposition 3.2.24, Proposition 3.1.7, and symmetry that $C \cap \neg\neg\neg S \bowtie C \cap \neg S$. Compute $r > 0$ such that if $x \in C \cap \neg\neg\neg S$ and $y \in C \cap \neg S$, then $\|x - y\| > r$. Assume that $S \times (C \cap \neg S)$ is inhabited by a point (x, y). Compute a positive integer N such that $\|x - y\| < Nr$, and for $0 \leqslant n \leqslant N$ define

$$t_n \equiv \frac{n}{N}, \quad x_n \equiv (1 - t_n)\, x + t_n y.$$

Then

$$x_0 = x \in S \subset C \cap \neg\neg\neg S,$$

and each $x_n \in C$. If $n < N$ and $x_n \in C \cap \neg\neg\neg S$, then since

$$\|x_{n+1} - x_n\| = \frac{1}{N}\|x - y\| < r,$$

our choice of r ensures that $x_{n+1} \notin C \cap \neg S$ and therefore $x_{n+1} \in C \cap \neg\neg\neg S$. It follows that $x_n \in C \cap \neg\neg\neg S$ for $0 \leqslant n \leqslant N$, so $y = x_N \in C \cap \neg\neg\neg S$, which is absurd. Hence, in fact, $S \times (C \cap \neg S) = \varnothing$, and therefore C is proximally connected. ∎

In particular, every interval in $\mathbf{R}$ is proximally connected. To see that the converse implies the law of excluded middle, let P be any statement, and take

$$X \equiv \{0, 1\} \cup \{x : 0 \leqslant x \leqslant 1 \wedge (P \vee \neg P)\}, \tag{3.43}$$

with the apartness induced by the standard metric apartness on $\mathbf{R}$. Let S be a subset of X such that $S \bowtie \neg S$, and suppose that $S \times \neg S$ is inhabited by an element (a, b). If $P \vee \neg P$, then $X = [0, 1]$, and so, by Proposition 3.8.1, $S \times \neg S = \varnothing$, a contradiction. It follows that $\neg (P \vee \neg P)$, which is absurd. We conclude that $S \times \neg S = \varnothing$, and hence that X is proximally connected. But if X is an interval, then $X = [0, 1]$ and therefore $P \vee \neg P$ holds.

In spite of the foregoing Brouwerian example, we have a weak converse to the proximal connectedness of intervals.

Proposition 3.8.2 *Let X be a proximally connected subset of $\mathbf{R}$, and let a, b be points of X with $a < b$. Then*

$$\forall_{x \in \mathbf{R}} \left(a < x < b \Rightarrow \forall_{\varepsilon > 0} \neg\neg \exists_{y \in (a,b) \cap X} \left(|x - y| < \varepsilon\right)\right).$$

Proof. Given x in (a, b) and $\varepsilon > 0$, we may assume that $[x - \varepsilon, x + \varepsilon] \subset (a, b)$. Suppose that

$$\neg \exists_{y \in (a,b) \cap X} \left(|x - y| < \varepsilon\right). \tag{3.44}$$

Then $|x - y| > \varepsilon/2$ for all y in $(a, b) \cap X$. Consider the set

$$S \equiv \left\{ s \in X : s < x - \frac{\varepsilon}{2} \right\}.$$

For each $y \in X \cap \neg S$ we have $y \geqslant x - \varepsilon/2$; if also $y < x + \varepsilon/2$, then $|x - y| \leqslant \varepsilon/2$, so $y \in (a, b) \cap X$ and therefore $|x - y| > \varepsilon/2$, which is absurd. It follows that $y \geqslant x + \varepsilon/2$; whence $\rho(y, s) > \varepsilon$ for each $s \in S$. Hence $S \bowtie (X \cap \neg S)$ in X, and therefore $S \times (X \cap \neg S) = \varnothing$. But this is absurd, as $a \in S$ and $b \in X \cap \neg S$. Hence the assumption (3.44) is false. ∎

As we would hope, proximal connectedness is preserved by the morphisms in the category of pre-apartness spaces:

Proposition 3.8.3 *Let f be a strongly continuous mapping of a proximally connected pre-apartness space X onto a pre-apartness space Y. Then Y is proximally connected.*

Proof. Let T be a subset of Y such that $T \bowtie \neg T$. Noting that $f^{-1}(\neg T) = \neg f^{-1}(T)$, we see from the strong continuity of f that $f^{-1}(T) \bowtie \neg f^{-1}(T)$. Since X is proximally connected, it follows that $f^{-1}(T) \times f^{-1}(\neg T) = \varnothing$; whence $T \times \neg T = \varnothing$. ∎

We can now prove a rather weak version of the intermediate value theorem.

Corollary 3.8.4 *If $f : [a, b] \to \mathbf{R}$ is strongly continuous and $f(a)f(b) < 0$, then*

$$\forall_{\varepsilon > 0} \neg\neg \exists_{x \in (a, b)} \left(|f(x)| < \varepsilon \right).$$

Proof. By Proposition 3.8.1, the interval $[a, b]$ is proximally connected. We can therefore apply Proposition 3.8.3 and then (noting that 0 is strictly between $f(a)$ and $f(b)$) Proposition 3.8.2. ∎

The downside of this weak version of the intermediate value theorem is that (i) the conclusion is a negative one and (ii) the proof depends on the strong continuity of f in order that we can apply Proposition 3.8.3. Using an approximate interval-halving argument like that in the proof of Theorem (2.5) on page 57 of [24], when $f(a)f(b) < 0$ we can derive the positive conclusion

$$\exists_{x \in (a, b)} \left(|f(x)| < \varepsilon \right)$$

under the much weaker assumption that f is sequentially continuous on $[a, b]$; but there seems no way to produce that positive conclusion with a proof of the kind used above for Proposition 3.8.2. On the other hand, to offset point (ii) we should bear in mind that Bishop [9] deals only with uniformly (and hence strongly) continuous maps on compact intervals; indeed, to produce an example of a continuous, but not uniformly continuous, mapping $f : [0, 1] \to \mathbf{R}$ we need to add a hypothesis such as the Church–Markov–Turing thesis; see [24] (Chapter 3).

Proposition 3.8.5 *Let A be a proximally connected subset of a symmetric pre-apartness space X that has the property* **A4$_{ss}$***, and let B be a subset of X such that $A \subset B \subset \overline{A}$ (where $\overline{A}$ denotes the closure of A in the apartness topology). Then B is proximally connected.*

Proof. Let $S \subset B$ satisfy $S \bowtie \neg S$, and suppose there exist $x \in S$ and $y \in B \cap \neg S$. Then $x \in \overline{A} \cap -\neg S$, so $A \cap -\neg S$ is inhabited. Also, since (by symmetry) $\neg S \bowtie S$, we see that

$$y \in B - S \subset \overline{A} - S;$$

whence $A - S$, and therefore $A \cap \neg S$, is inhabited. On the other hand, by Proposition 3.1.7, $\neg S \bowtie \neg \neg S$. Since $-\neg S \subset \neg \neg S$, it follows that

$$(A \cap \neg S) \bowtie (A \cap -\neg S)$$

in the proximally connected space A. Hence

$$(A \cap \neg S) \times (A \cap -\neg S) = \varnothing,$$

which contradicts what we showed earlier in the proof. We conclude that $S \times \neg S = \varnothing$. ∎

Corollary 3.8.6 *The closure (in the apartness topology) of a proximally connected subset of a symmetric pre-apartness space that has the property* **A4$_{ss}$** *is proximally connected.*

Proposition 3.8.7 *Let X be a pre-apartness space that is the union of a family $(X_i)_{i \in I}$ of proximally connected subspaces such that $\bigcap_{i \in I} X_i$ is inhabited. Then X is proximally connected.*

Proof. Fix $\xi \in \bigcap_{i \in I} X_i$. Let S be a subset of X such that $S \bowtie \neg S$, and suppose that $S \times \neg S$ contains a point (a, b). Pick $i_0 \in I$ such that $a \in X_{i_0}$. Noting that

$$(X_{i_0} \cap S) \bowtie (X_{i_0} \cap \neg S),$$

that X_{i_0} is proximally connected, and that $a \in X_{i_0} \cap S$, we obtain $X_{i_0} \cap \neg S = \varnothing$. Since $\xi \in X_{i_0}$, it follows that $\xi \in \neg \neg S$. On the other hand, choosing i_1 such that $b \in X_{i_1}$, and using a similar argument, we can show that $X_{i_1} \cap S = \varnothing$ and hence that $\xi \in \neg S$, which is absurd. Hence $S \times \neg S = \varnothing$. ∎

To deal with proximal connectedness in product apartness spaces, we need a couple of lemmas.

Lemma 3.8.8 *Let X_1, X_2 be symmetric apartness spaces, let $x_1 \in X_1$, and let S, T be subsets of $\{x_1\} \times X_2$. Then $S \bowtie T$ in $\{x_1\} \times X_2$ if and only if $\mathrm{pr}_2(S) \bowtie \mathrm{pr}_2(T)$ in X_2.*

Proof. From the definition of apartness in the product space $\{x_1\} \times X_2$, we see that $S \bowtie T$ in $\{x_1\} \times X_2$ if and only if there exist subsets $A_1,\ldots,A_m$ and $B_1,\ldots,B_n$ of X_2 such that

$$S \subset \bigcup_{i=1}^{m} (\{x_1\} \times A_i), \quad T \subset \bigcup_{j=1}^{n} (\{x_1\} \times B_j)$$

and

$$A_i \bowtie B_j \quad (1 \leqslant i \leqslant m \wedge 1 \leqslant j \leqslant n).$$

This last condition is equivalent to

$$\bigcup_{i=1}^{m} A_i \bowtie \bigcup_{j=1}^{n} B_j. \tag{3.45}$$

Without loss of generality, we can replace A_i by $A_i \cap \mathrm{pr}_2(S)$ and B_j by $B_j \cap \mathrm{pr}_2(T)$. We then see that (3.45) is equivalent to $\mathrm{pr}_2(S) \bowtie_{X_2} \mathrm{pr}_2(T)$. ∎

The second lemma is easily derived either from the first or using an analogous proof.

Lemma 3.8.9 *Let X_1,X_2 be symmetric apartness spaces, let $x_2 \in X_2$, and let S,T be subsets of $X_1 \times \{x_2\}$. Then $S \bowtie T$ in $X_1 \times \{x_2\}$ if and only if $\mathrm{pr}_1(S) \bowtie \mathrm{pr}_1(T)$ in X.*

Proposition 3.8.10 *Let X_1,X_2 be symmetric apartness spaces. Then $X_1 \times X_2$ is proximally connected if and only if both X_1 and X_2 are proximally connected.*

Proof. If $X \equiv X_1 \times X_2$ is proximally connected, then, since (by Corollary 3.7.3) the projection maps are strongly continuous, both X_1 and X_2 are proximally connected, by Proposition 3.8.3.

Conversely, suppose that X_1 and X_2 are proximally connected. Given $x \in X_1$ and $y \in X_2$, let S be a subset of $\{x\} \times X_2$ such that

$$S \bowtie ((\{x\} \times X_2) \cap \neg S) \text{ in } \{x\} \times X_2.$$

By Lemma 3.8.8,

$$\mathrm{pr}_2(S) \bowtie \mathrm{pr}_2((\{x\} \times X_2) \cap \neg S) \text{ in } X_2.$$

But

$$\neg\mathrm{pr}_2(S) = \mathrm{pr}_2((\{x\} \times X_2) \cap \neg S),$$

so $\mathrm{pr}_2(S) \bowtie \neg\mathrm{pr}_2(S)$ in X_2. Since X_2 is proximally connected, it follows that

$$\mathrm{pr}_2(S) \times \mathrm{pr}_2((\{x\} \times X_2) \cap \neg S) = \mathrm{pr}_2(S) \times \neg\mathrm{pr}_2(S) = \varnothing;$$

whence

$$S \times ((\{x\} \times X_2) \cap \neg S) = \varnothing.$$

Thus the space $\{x\} \times X_2$ is proximally connected. Similarly, using Lemma 3.8.9, we prove that $X_1 \times \{y\}$ is proximally connected.

Now fix $x_1 \in X_1$ and $x_2 \in X_2$. By the first part of this proof, for any $x \in X_1$ the set $\{x\} \times X_2$ is proximally connected, as is the set $X_1 \times \{x_2\}$. Since each of these two sets contains (x, x_2), we see from Proposition 3.8.7 that

$$C_x \equiv (\{x\} \times X_2) \cup (X_1 \times \{x_2\})$$

is proximally connected. For each $x \in X_1$ the set C_x contains (x_1, x_2), so another application of Proposition 3.8.7 shows that

$$X = \bigcup_{x \in X_1} C_x$$

is proximally connected. ∎

For information about the relations between our proximal connectedness and other types of connectedness in constructive mathematics, see [30].

3.9 Towards Nearness

Let $(X, \bowtie)$ be a symmetric apartness space. In this section we investigate further the possibility of finding uniform structures that induce the given apartness on X. This will lead us to a notion of proximity, or nearness, for subsets of X, something we have paid scant attention to despite its being the primary notion in the classical counterpart of our theory.

Given inhabited subsets A and B of X, we say that A is **well contained** in B, and we write $A \ll B$, if there exists $C \subset X$ such that $X = B \cup C$ and $A \bowtie C$; in that case, $A \subset B$. Let $\mathcal{B}_w$ be the class of all sets of the form

$$\bigcup_{i=1}^{m} (B_i \times B_i)$$

where there exist finitely many subsets $A_1, \ldots, A_m$ of X such that $A_i \ll B_i$ for each i and $X = \bigcup_{i=1}^{m} A_i$. The class $\mathcal{B}_w$ is inhabited: since $X = X \cup \varnothing$ and $X \bowtie \varnothing$, we have $X \times X \in \mathcal{B}_w$.

The **apartness class of uniformities** for X is the set $\mathcal{A}_{X,\bowtie}$ of uniform structures $\mathcal{U}$ for which the corresponding apartness $\bowtie_\mathcal{U}$ is precisely the given apartness $\bowtie$ on X. Classically, if X has the Efremovič property, then $\mathcal{A}_{X,\bowtie}$ is inhabited and contains a unique totally bounded member, for which $\mathcal{B}_w$ is a basis of entourages ([75], page 72, (12.3)). Constructively, we cannot prove that $\mathcal{A}_{X,\bowtie}$ is inhabited in general. Nevertheless, as we aim to show, under reasonable conditions on the original apartness on X we can prove that

$$S \bowtie_w T \Leftrightarrow \exists_{U \in \mathcal{B}_w} (S \times T \subset \neg U) \tag{3.46}$$

defines a set-set apartness on X.

Proposition 3.9.1 *Let X be a symmetric apartness space. Then each element of $\mathcal{B}_w$ is symmetric and contains the diagonal Δ of $X \times X$. Moreover, $\mathcal{B}_w$ is closed under finite intersections.*

Proof. It is clear that each element of $\mathcal{B}_w$ is symmetric and contains the diagonal; it will suffice, then, to prove that the intersection of two elements of $\mathcal{B}_w$ is also in $\mathcal{B}_w$. Let $A_1, \ldots, A_m, B_1, \ldots, B_m$ be subsets of X such that $A_i \ll B_i$ for each i and $X = \bigcup_{i=1}^{m} A_i$; and let $P_1, \ldots, P_n, Q_1, \ldots, Q_n$ be subsets of X such that $P_j \ll Q_j$ for each j and $X = \bigcup_{j=1}^{n} P_j$. Then the sets

$$S \equiv \bigcup_{i=1}^{m} (B_i \times B_i) \text{ and } T \equiv \bigcup_{j=1}^{n} (Q_j \times Q_j)$$

belong to $\mathcal{B}_w$. For each i $(1 \leqslant i \leqslant m)$ choose C_i such that $X = B_i \cup C_i$ and $A_i \bowtie C_i$; and for each j $(1 \leqslant j \leqslant n)$ choose R_j such that $X = Q_j \cup R_j$ and $P_j \bowtie R_j$. Then

$$X = (B_i \cap Q_j) \cup L_{ij},$$

where

$$L_{ij} \equiv (B_i \cap R_j) \cup (C_i \cap Q_j) \cup (C_i \cap R_j)$$

and

$$A_i \cap P_j \bowtie L_{ij}.$$

On the other hand,

$$X = \left(\bigcup_{i=1}^{m} A_i\right) \cap \left(\bigcup_{j=1}^{n} P_j\right) = \bigcup_{i=1}^{m} \bigcup_{j=1}^{n} (A_i \cap P_j).$$

Hence

$$S \cap T = \bigcup_{i=1}^{m} \bigcup_{j=1}^{n} ((B_i \cap Q_j) \times (B_i \cap Q_j))$$

belongs to $\mathcal{B}_w$. ∎

Proposition 3.9.2 *Let X be a symmetric apartness space, let x, y be elements of X, and let U be an element of $\mathcal{B}_w$ such that $(x, y) \in \neg U$. Then $x \neq y$.*

Proof. There exist finitely many subsets A_i, B_i $(1 \leqslant i \leqslant m)$ of X such that $A_i \ll B_i$ for each i, $X = \bigcup_{i=1}^{m} A_i$, and $U = \bigcup_{i=1}^{m} (B_i \times B_i)$. Pick i such that $y \in A_i \subset B_i$, and then choose $C_i \subset X$ such that $X = B_i \cup C_i$ and $A_i \bowtie C_i$. Since $(x, y) \notin U$, x cannot belong to B_i, so $x \in C_i \subset -A_i \subset {\sim}A_i$ and therefore $x \neq y$. ∎

Proposition 3.9.3 *Let X be a symmetric apartness space. Then the relation $\bowtie_w$, defined at (3.46), is symmetric and satisfies* **B1–B3** *and the condition*

$$\forall_{x \in X} (x \bowtie_w B \Rightarrow x \bowtie B). \tag{3.47}$$

Moreover,

$$\forall_{A, B \subset X} (A \bowtie_w B \Rightarrow A \bowtie B) \tag{3.48}$$

if and only if

$$\forall_{A, B \subset X} (A \times B = \varnothing \Rightarrow A \bowtie B). \tag{3.49}$$

Proof. Axiom **B1** holds for $\bowtie_w$ because $X \times \varnothing \subset \neg U$ for any $U \in \mathcal{B}_w$. If $x \bowtie_w S$, then there exists $U \in \mathcal{B}_w$ such that $\{x\} \times S \subset \neg U$, from which, with reference to Proposition 3.9.2, we obtain $x \in {\sim}S$; thus **B2** holds.

If $S_1 \bowtie_w T_1$ and $S_1 \bowtie_w T_2$, and we choose $U_k \in \mathcal{B}_w$ such that $S_1 \times T_k \subset \neg U_k$, then

$$S_1 \times (T_1 \cup T_2) = (S_1 \times T_1) \cup (S_1 \times T_2) \subset \neg U_1 \cup \neg U_2 \subset \neg (U_1 \cap U_2),$$

where, by Proposition 3.9.1, $U_1 \cap U_2 \in \mathcal{B}_w$. Hence $S_1 \bowtie_w (T_1 \cup T_2)$. If, on the other hand, $S_1 \bowtie_w (T_1 \cup T_2)$ and we choose $U \in \mathcal{B}_w$ such that $S_1 \times (T_1 \cup T_2) \subset \neg U$, then we have $S_1 \times T_1 \subset \neg U$ and $S_1 \times T_2 \subset \neg U$; whence $S_1 \bowtie_w T_1$ and $S_1 \bowtie_w T_2$. An application of the symmetry of $\bowtie_w$, which clearly follows from that of the elements of $\mathcal{B}_w$, completes the verification of **B3**.

Next, if (3.48) holds and $A \times B = \varnothing = \neg (X \times X)$, then $A \bowtie_w B$ and therefore $A \bowtie B$. Thus (3.48) entails (3.49).

Now let $A \bowtie_w B$, and choose subsets $S_1, \ldots, S_m, T_1, \ldots, T_m, C_1, \ldots, C_m$ of X such that $S_i \bowtie C_i$ and $X = T_i \cup C_i$ for each i, $X = \bigcup\limits_{i=1}^{m} S_i$, and

$$A \times B \subset \neg \bigcup_{i=1}^{m} (T_i \times T_i) = \bigcap_{i=1}^{m} \neg (T_i \times T_i). \tag{3.50}$$

For the moment, fix $i \leqslant m$. Note that

$$A \cap C_i \bowtie B \cap S_i. \tag{3.51}$$

On the other hand, we see from (3.50) that

$$(A \cap T_i) \times (B \cap T_i) = \varnothing, \tag{3.52}$$

so if (3.49) holds, then

$$A \cap T_i \bowtie B \cap T_i \supset B \cap S_i. \tag{3.53}$$

If $A \equiv \{x\}$ is a singleton and $x \in T_i$, then $B \cap T_i = \varnothing$, by (3.52), so (3.53) holds without our assuming (3.49). Taking (3.51) and (3.53) together, we see that if either (3.49) holds or A is a singleton, then

$$A = (A \cap T_i) \cup (A \cap C_i) \bowtie B \cap S_i.$$

Applying **B3** then yields

$$A \bowtie \bigcup_{i=1}^{m} (B \cap S_i) = B \cap \bigcup_{i=1}^{m} S_i = B,$$

which completes the proof of (3.47) in the general case and, when (3.49) is assumed, that of (3.48). ∎

Property (3.49) is a strengthening of axiom **B1**. To see that it holds in a uniform apartness space $(X, \mathcal{U})$, let A, B be subsets of X with $A \times B = \varnothing$; then for each $U \in \mathcal{U}$ we have $A \times B \subset \neg U$, so $A \bowtie B$.

Does every symmetric apartness space satisfy (3.49)? To answer this negatively, let X be an inhabited set with the denial inequality, and consider the apartness $\bowtie_\varnothing$ defined on X by

$$A \bowtie_\varnothing B \Leftrightarrow (A = \varnothing \vee B = \varnothing).$$

By Corollary 3.2.30, if $\bowtie_\varnothing$ is induced by a uniform structure, then we can derive the weak law of excluded middle. Fixing an element a of X, define

$$A \equiv \{x \in X : x = a \wedge P\}, \quad B \equiv \{x \in X : x = a \wedge \neg P\}. \quad (3.54)$$

Then $A \times B = \varnothing$; but if $A \bowtie_\varnothing B$, then either $A = \varnothing$ and $\neg P$ holds, or else $B = \varnothing$ and $\neg\neg P$ holds. Thus if this apartness satisfies (3.49)—or, equivalently, (3.48)—then we can derive the weak law of excluded middle.

We now re-introduce the Efremovič property into our deliberations.

Proposition 3.9.4 *If $A \ll B$ in a symmetric apartness space X that has the Efremovič property, then there exists U such that $A \subset -U$ and $-U \ll B$.*

Proof. There exists $C \subset X$ such that $X = B \cup C$ and $A \bowtie C$. Choose E such that $A \bowtie \neg E$ and $E \bowtie C$, and set $U \equiv \neg E$. Then $A \subset -U$. On the other hand, by symmetry and Propositions 3.1.10 and 3.1.7, $\neg\neg E \bowtie C$. Since $-\neg E \subset \neg\neg E$, it follows that $-U \bowtie C$; whence $-U \ll B$. ∎

It follows that in the presence of symmetry and the Efremovič condition, $\mathcal{B}_w$ is precisely the class of all sets of the form

$$\bigcup_{i=1}^{n} (B_i \times B_i)$$

where there exist apartness complements $-U_1, \ldots, -U_n$ of X such that $-U_i \ll B_i$ for each i and $X = \bigcup_{i=1}^{n} -U_i$.

Lemma 3.9.5 *Let X be a symmetric apartness space with the Efremovič property, and let $x \in -A$. Then $A \ll -\{x\}$, and there exists W such that $x \in -W$ and $\neg W \ll -A$.*

Proof. There exists E such that $x \bowtie \neg E$ and $E \bowtie A$. Then $x \in -\neg E$, so, by Lemma 3.1.5,

$$X = -\{x\} \cup -\neg E = -\{x\} \cup \neg\neg E.$$

Since also $A \bowtie \neg\neg E$ by symmetry and Propositions 3.1.10 and 3.1.7, we see that $A \ll -\{x\}$. Now, by **B5**, there exists S such that $x \in -S$ and $X = -A \cup S$. Applying the Efremovič property once more, we obtain F such that $x \bowtie \neg F$ and $F \bowtie S$. Then (see above, and note that symmetry is again used) $\neg\neg F \bowtie S$. We complete the proof by taking $W \equiv \neg F$. ∎

Proposition 3.9.6 *Let X be a symmetric apartness space, and $x \in X$. Then for each $U \in \mathcal{B}_w$, there exists $S \subset X$ such that $x \in -S \subset U[x]$.*

Proof. Let $U \equiv \bigcup_{i=1}^{n} (B_i \times B_i)$, where there exist subsets $A_1,\ldots,A_n$ of X such that $A_i \ll B_i$ $(1 \leqslant i \leqslant n)$ and $X = \bigcup_{i=1}^{n} A_i$. For each i, there exists $C_i \subset X$ such that $X = B_i \cup C_i$ and $A_i \bowtie C_i$. Choose i such that $x \in A_i$; then $x \in -C_i$. If $y \in -C_i$, then $y \in B_i$; whence

$$(x, y) \in A_i \times B_i \subset B_i \times B_i \subset U$$

and therefore $y \in U[x]$. Hence $x \in -C_i \subset U[x]$, so we can take $S \equiv C_i$. ∎

Proposition 3.9.7 *Let X be a symmetric apartness space with the Efremovič property, and let $x \in -A$. Then $x \in U[x] \subset -A$, where*

$$U \equiv (-\{x\} \times -\{x\}) \cup (-A \times -A) \in \mathcal{B}_w.$$

Proof. First use Lemma 3.9.5 to construct W such that $x \in -W$ and $\neg W \ll -A$. Then use **B5** to construct V such that $x \in -V$ and $X = -W \cup V$. By the first part of Lemma 3.9.5, $V \ll -\{x\}$. Since $V \cup \neg W = X$, we now see that $U \in \mathcal{B}_w$. By Proposition 3.9.1, $(x, x) \in U$ and therefore $x \in U[x]$. Given $y \in U[x]$, we have $(x, y) \in U$ and therefore, clearly, $(x, y) \in -A \times -A$; whence $y \in -A$. Thus $x \in U[x] \subset -A$. ∎

Corollary 3.9.8 *Let $(X, \bowtie)$ be a symmetric apartness space with the Efremovič property. Then the sets of the form $U[x]$, with $x \in X$ and $U \in \mathcal{B}_w$, constitute a base for the apartness topology on X.*

Proof. This follows from Propositions 3.9.6 and 3.9.7. ∎

Proposition 3.9.6 also leads to a converse of Proposition 3.9.2.

Corollary 3.9.9 *Let X be a symmetric, $\mathbf{T}_1$ apartness space with the Efremovič property, and let $x \neq y$ in X. Then there exists $U \in \mathcal{B}_w$ such that $(x, y) \in \neg U$.*

Proof. Since X is $\mathbf{T}_1$, we have $x \in -\{y\}$. Hence, by Proposition 3.9.7,

$$U \equiv (-\{x\} \times -\{x\}) \cup (-\{y\} \times -\{y\})$$

belongs to $\mathcal{B}_w$. Clearly, $(x,y) \in \neg U$. ∎

Let $(X, \mathcal{U})$ be a uniform space, and τ a topology on X. We say that $\mathcal{U}$ is **compatible with the topology** τ if the corresponding uniform topology coincides with τ. On the other hand, we say that a set S of subsets of $X \times X$ **generates** the uniform structure $\mathcal{U}$ if each member of S is in $\mathcal{U}$, and each member of $\mathcal{U}$ contains a member of S.

Corollary 3.9.10 *Let $(X, \bowtie)$ be a symmetric apartness space with the Efremovič property. If $\mathcal{B}_w$ generates a uniform structure $\mathcal{U}_w$, then $\mathcal{U}_w$ is compatible with the apartness topology on X.*

Proof. This follows from Corollary 3.9.8 and the fact that the sets $U[x]$, with $x \in X$ and $U \in \mathcal{B}_w$, form a base for the topology $\tau_{\mathcal{U}_w}$ (by definition of the latter). ∎

We can now give conditions under which the point-set apartness $\bowtie_w$ (see (3.46)) coincides with $\bowtie$.

Proposition 3.9.11 *If a symmetric apartness space $(X, \bowtie)$ has the Efremovič property, then*

$$\forall_{x \in X} \forall_A (x \bowtie A \Leftrightarrow x \bowtie_w A).$$

Proof. In view of Proposition 3.9.3, we need only prove the left-to-right implication. Suppose, then, that $x \bowtie A$ and therefore $x \in -A$. By Proposition 3.9.7, $x \in U[x] \subset -A$, where

$$U \equiv (-\{x\} \times -\{x\}) \cup (-A \times -A) \in \mathcal{B}_w.$$

To show that $x \bowtie_w A$, it is enough to prove that $\{x\} \times A \subset \neg U$. For $y \in A$ and $\mathbf{u} \in U$, if $\mathbf{u} \in -\{x\} \times -\{x\}$, then $u_1 \neq x$ and so $(x,y) \neq \mathbf{u}$; whereas if $\mathbf{u} \in -A \times -A$, then $u_2 \in -A \subset \neg A$, so $u_2 \neq y$ and hence $(x,y) \neq \mathbf{u}$. ∎

We see from Corollary 3.9.8 and Proposition 3.9.11 that, under the Efremovič condition on a symmetric apartness space X, even if $\bowtie_w$ is not a set-set apartness on X—and hence even if $\mathcal{B}_w$ does not generate a uniform structure (let alone one that induces the apartness $\bowtie$)—the relations $\bowtie_w$ and $\mathcal{B}_w$ are nicely linked with the apartness $\bowtie$ and the topology $\tau_\bowtie$ on X.

Corollary 3.9.12 *If a symmetric apartness space $(X, \bowtie)$ has the Efremovič property, then the set-set relation $\bowtie_w$ satisfies axioms **B1–B5**, and the corresponding point-set apartness coincides with the point-set apartness induced by $\bowtie$.*

Proof. By Proposition 3.9.3, $\bowtie_w$ satisfies **B1–B3**. On the other hand, since, by Proposition 3.9.11, the point-set relations corresponding to $\bowtie$ and $\bowtie_w$ coincide, the latter satisfies **B5** and hence **B4**. ∎

We now explore the connection between $\mathcal{B}_w$ and uniform structures that induce the apartness on X. For this exploration, we say that $\mathcal{B}_w$ **strongly generates** the uniform structure $\mathcal{U}$ on X if for each $U \in \mathcal{U}$, there exist *inhabited* subsets A_i, B_i $(1 \leqslant i \leqslant n)$ of X such that $X = \bigcup_{i=1}^{n} A_i$, $A_i \ll B_i$ for each i, and $\bigcup_{i=1}^{n} (B_i \times B_i) \subset U$.

Proposition 3.9.13 *Let* $(X, \mathcal{U})$ *be a uniform apartness space, and define* $\mathcal{B}_w$ *relative to the uniform apartness on* X *as at the start of this section. Then for each* $V \in \mathcal{B}_w$, *there exists* $U \in \mathcal{U}$ *such that* $U \subset V$. *If also* $\mathcal{B}_w$ *strongly generates* $\mathcal{U}$, *then* $(X, \mathcal{U})$ *is totally bounded.*

Proof. Let A_i, B_i, C_i $(1 \leqslant i \leqslant n)$ be subsets of X such that $X = \bigcup_{i=1}^{n} A_i$ and for each i, $X = B_i \cup C_i$ and $A_i \bowtie C_i$. Choose $U \in \mathcal{U}$ such that $A_i \times C_i \subset \neg U$ for each i. Let $(x, y) \in U$, and pick i with $x \in A_i \subset B_i$. Then $y \notin C_i$, so $y \in B_i$ and therefore $(x, y) \in B_i \times B_i$. Hence $U \subset \bigcup_{i=1}^{n} (B_i \times B_i)$. This completes the proof of the first part of the proposition.

Suppose now that $\mathcal{B}_w$ strongly generates $\mathcal{U}$, and consider a symmetric entourage U in $\mathcal{U}$. Construct sets A_i, B_i, C_i as in the first part of the proof, such that also each A_i is inhabited and $\bigcup_{i=1}^{n} (B_i \times B_i) \subset U$. For each i pick x_i in A_i. For each $x \in B_i$ we have $(x_i, x) \in B_i \times B_i \subset U$, so $x \in U[x_i]$. Hence $B_i \subset U[x_i]$ and therefore $A_i \subset U[x_i]$. It follows that $\{x_1, \ldots, x_n\}$ is a finitely enumerable U-approximation to X. ∎

Proposition 3.9.14 *Let* $(X, \mathcal{U})$ *be a totally bounded uniform apartness space. Then* $\mathcal{U}$ *is strongly generated by* $\mathcal{B}_w$.

Proof. Given U_1 in $\mathcal{U}$, construct a 4-chain (U_1, U_2, U_3, U_4) of entourages of X with U_2 symmetric. Let $\{x_1, \ldots, x_n\}$ be a U_4-approximation to X, and set

$$A_i \equiv U_4[x_i], \quad B_i \equiv U_2[x_i], \quad C_i \equiv \neg U_3[x_i].$$

Then $x_i \in A_i$ and $X = B_i \cup C_i$. If $(x, y) \in A_i \times C_i$ then $(x_i, x) \in U_4$ and $(x_i, y) \in \neg U_3$; if also $(x, y) \in U_4$, then $(x_i, y) \in U_4^2 \subset U_3$, a contradiction. Thus $A_i \times C_i \subset \neg U_4$ and therefore $A_i \bowtie C_i$. It now follows that $A_i \ll B_i$ and hence that $\bigcup_{i=1}^{n} (B_i \times B_i) \in \mathcal{B}_w$. Moreover, if $(z, z') \in B_i \times B_i$, then $(z, x_i) \in$

U_2 (recall that U_2 is symmetric) and $(x_i, z') \in U_2$, so $(z, z') \in U_2^2 \subset U_1$. Thus $\bigcup_{i=1}^{n} (B_i \times B_i) \subset U_1$. ∎

Corollary 3.9.15 *Let $(X, \bowtie)$ be a symmetric apartness space. Then there is at most one uniform structure on X that is totally bounded and induces the given apartness.*

Proof. Suppose there exists a totally bounded uniform structure $\mathcal{U}$ in $\mathcal{A}_{X,\bowtie}$. By Proposition 3.9.14, $\mathcal{U}$ is (strongly) generated by $\mathcal{B}_w$ and so is unique. ∎

Corollary 3.9.16 *Let $(X, \bowtie)$ be a symmetric apartness space such that $\mathcal{A}_{X,\bowtie}$ contains a totally bounded member T. Then $T \subset \mathcal{U}$ for each $\mathcal{U} \in \mathcal{A}_{X,\bowtie}$.*

Proof. Let $\mathcal{U} \in \mathcal{A}_{X,\bowtie}$ and $T \in \mathcal{T}$. By Proposition 3.9.14, there exist $A_i, B_i \subset X$ $(1 \leqslant i \leqslant n)$ such that $X = \bigcup_{i=1}^{n} A_i$, $A_i \ll B_i$ for each i, and

$$V \equiv \bigcup_{i=1}^{n} (B_i \times B_i) \subset T.$$

For each i choose $C_i \subset X$ such that $X = B_i \cup C_i$ and $A_i \bowtie C_i$. Since $\mathcal{U} \in \mathcal{A}_{X,\bowtie}$, for each i we can find $U_i \in \mathcal{U}$ such that

$$A_i \times C_i \subset \neg U_i.$$

Writing

$$U \equiv \bigcap_{i=1}^{n} U_i,$$

we see that $U \in \mathcal{U}$ and that

$$A_i \times C_i \subset \bigcup_{i=1}^{n} \neg U_i \subset \neg U \quad (1 \leqslant i \leqslant n).$$

For each $(x, y) \in U$, there exists i such that $x \in A_i$. Then $y \notin C_i$, so $y \in B_i$ and therefore $(x, y) \in B_i \times B_i$. Hence $U \subset V \subset T$ and therefore $T \in \mathcal{U}$. ∎

For subsets S, T of any symmetric apartness space $(X, \bowtie)$, Proposition 3.9.3 shows that if (3.49) holds, then $S \bowtie_w T$ implies $S \bowtie T$. Can we prove constructively that, as classically, if (3.49) holds, then the set-set apartness relations $\bowtie$ and $\bowtie_w$ coincide? If X is a totally bounded uniform apartness space, the answer is "yes".

Proposition 3.9.17 *If $(X, \mathcal{U})$ is a totally bounded uniform space, then*

$$\forall_{S,T \subset X} (S \bowtie T \Leftrightarrow S \bowtie_w T).$$

Proof. If $S \bowtie T$, then there exists $U \in \mathcal{U}$ such that $S \times T \subset \neg U$. By Proposition 3.9.14, there exists $V \in \mathcal{B}_w$ such that $V \subset U$ and therefore $S \times T \subset \neg V$. Hence $S \bowtie_w T$. Since every uniform space satisfies (3.49), reference to Proposition 3.9.3 completes the proof. ∎

Although we have discussed the notion of a point being near a set relative to a pre-apartness relation, we have not yet considered the nearness/proximity of two sets. Classically, if a proximity space X has the Efremovič property, then $\mathcal{B}_w$ generates the unique totally bounded uniform structure $\mathcal{U}_w$ compatible with the given apartness on X. In that case, in which *near* is the denial of *apart*, subsets S,T of X are near if and only if $S \times T$ intersects each $U \in \mathcal{U}_w$; from which it follows that S and T are near if and only if $S \times T$ intersects each element of $\mathcal{B}_w$. This suggests our constructive definition: for subsets S,T of a symmetric set-set apartness space X, we say that S is **near** T, and we write $\delta(S,T)$, if $S \times T$ intersects each $U \in \mathcal{B}_w$. In that case, by the symmetry of $\mathcal{B}_w$, T is near S.

How does this definition square with the definition of nearness for a point and a set, namely

$$\text{near}\,(x, A) \Leftrightarrow \forall_{U \subset X}\,(x \in -U \Rightarrow \exists_y\,(y \in A - U))\,,$$

given in Section 2.1?

Proposition 3.9.18 *Let X be a symmetric apartness space with the Efremovič property, let $x \in X$, and let $A \subset X$. Then* near (x, A) *if and only if $\delta(x, A)$.*

Proof. Suppose first that $\delta(x, A)$. Consider any S such that $x \in -S$. By Proposition 3.9.7, the set

$$U \equiv (-\{x\} \times -\{x\}) \cup (-S \times -S)$$

belongs to $\mathcal{B}_w$. By definition of the relation δ, there exists $y \in A$ such that $(x, y) \in U$. Clearly, $(x, y) \in -S \times -S$, so $y \in -S$ and therefore $y \in A - S$. Since S is arbitrary, it follows that near(x, A).

Now suppose, conversely, that near(x, A), and let $U \in \mathcal{B}_w$. By Proposition 3.9.6, there exists $S \subset X$ such that $x \in -S \subset U[x]$. By definition of *near* there exists $y \in A - S \subset U[x]$; whence $(x, y) \in (\{x\} \times A) \cap U$. Since U is arbitrary, we conclude that $\delta(x,A)$. ∎

Thus, in the presence of symmetry and the Efremovič property, we have a set-set relation δ that is classically equivalent to proximity—denial of apartness—and that, for a point and a set, is constructively equivalent to the notion of point-set nearness used in our earlier work. Moreover, when X is a totally bounded uniform apartness space, it follows from Propositions 3.9.13 and 3.9.14 that

$$\forall_{S,T \subset X}\,(\delta(S,T) \Leftrightarrow \forall_{U \in \mathcal{U}}\,((S \times T) \cap U \neq \varnothing)\,,$$

so δ coincides with the usual classical notion of proximity in a uniform space. In view of all this, it seems that δ is an appropriate binary nearness predicate for further investigation in the constructive setting.

3.10   Neat Compactness

In this final section of the book we address the outstanding question: how can we define compactness for general, not necessarily uniform, apartness spaces? We follow Diener's path [39] to a notion of compactness via so-called *neat covers* and *neat completeness*.

Let $(X, \bowtie)$ be a symmetric pre-apartness space. We say that an ordered pair (S, T) of subsets of X is a **neat cover** of X if there exist subsets S', T' of X such that

$$S' \bowtie T', \ X = S \cup S', \text{ and } X = T \cup T'. \tag{3.55}$$

For example, if X is a metric space, $0 \leqslant \alpha < \beta$, and

$$\begin{aligned} S &\equiv \{x \in X : \rho(\xi, x) > \alpha\}, \\ T &\equiv \{x \in X : \rho(\xi, x) < \beta\}, \end{aligned}$$

then (S, T) is a neat cover of X, since the requirements of the definition are fulfilled by taking

$$S' \equiv \left\{x \in X : \rho(\xi, x) < \alpha + \frac{1}{3}(\beta - \alpha)\right\},$$

$$T' \equiv \left\{x \in X : \rho(\xi, x) > \beta - \frac{1}{3}(\beta - \alpha)\right\}.$$

Another example of a neat cover is given by the second of the following lemmas.

Lemma 3.10.1 *Let $(X, \mathcal{U})$ be a uniform apartness space, and (U, V) a 2-chain of entourages of X with V symmetric. Then $\neg U[x] \bowtie V[x]$ for each $x \in X$.*

Proof. Given $y \in \neg U[x]$ and $z \in V[x]$, suppose that $(y, z) \in V$. Then $(x, y) \in V \circ V^{-1} = V^2 \subset U$; so $y \in U[x]$, a contradiction. It follows that $\neg U[x] \times V[x] \subset \neg V$ and hence that $\neg U[x] \bowtie V[x]$. ∎

Lemma 3.10.2 *Let $(X, \mathcal{U})$ be a uniform space, and (U_1, U_2, U_3, U_4) a 4-chain of symmetric entourages of X. Then for each $x \in X$, the pair $(U_1[x], \neg U_4[x])$ is a neat cover of X.*

Proof. The result follows from the observations that

$$X = U_1[x] \cup \neg U_2[x] = \neg U_4[x] \cup U_3[x].$$

and that (by Lemma 3.10.1), $\neg U_2[x] \bowtie U_3[x]$. ∎

A subset A of a pre-apartness space X is said to be **neatly located** (in X) if for each neat cover (S,T) of X, either $A \subset S$ or $A \cap T$ is inhabited.

Proposition 3.10.3 *Every neatly located subset of a metric space is located.*

Proof. Let A be a neatly located subset of the metric space (X,ρ), and fix $\xi \in X$. Given α,β with $0 \leqslant \alpha < \beta$, let (S,T) be the neat cover of X introduced immediately before the definition of *neatly located*. Then either $A \subset S$ or $A \cap T$ is inhabited. In the first case, $\rho(\xi,x) > \alpha$ for all $x \in A$; in the second, there exists $x \in A$ with $\rho(\xi,x) < \beta$. Hence $\rho(\xi,A)$ exists, by the least-upper-bound principle (page 7). ∎

We shall see, in Proposition 3.10.5 below, that the converse of Proposition 3.10.3 does not hold constructively. First we prove

Proposition 3.10.4 *Every totally bounded subset of a uniform apartness space is neatly located.*

Proof. Let A be a totally bounded subset of the uniform space $(X,\mathcal{U})$, and let (S,T) be a neat cover of X. Construct subsets S',T' of X such that (3.55) holds. In view of Proposition 3.2.11, there exists a symmetric entourage U of X such that $S' \times T' \subset \neg U$. Construct a finite U-approximation $\{x_1, \ldots, x_m\}$ to A. Either there exists k such that $x_k \in T$, or else $x_i \in T'$ for each i $(1 \leqslant i \leqslant m)$. In the latter case, given $x \in A$ and using symmetry, choose i such that $(x, x_i) \in U$. If $x \in S'$, then our choice of U yields $(x, x_i) \in \neg U$, a contradiction. It follows that $A \subset \neg S' \subset S$. ∎

Proposition 3.10.5 *The subspace $[0,\infty)$ is neatly located in the metric apartness space* $\mathbf{R}$ *if and only if* **LPO** *holds.*

Proof. Let $(a_n)_{n \geqslant 1}$ be a decreasing binary sequence with $a_1 = 1$, and define

$$S \equiv \bigcup_{n \geqslant 1} (-\infty, na_n],$$

$$T \equiv \bigcap \{[(n-1)a_n, \infty) : a_{n+1} = 1 - a_n\}.$$

Define also

$$S' \equiv \bigcap_{n \geqslant 1} \left\{\left[\left(n - \frac{1}{3}\right) a_n, \infty\right) : a_{n+1} = 1 - a_n\right\},$$

$$T' \equiv \bigcup_{n \geqslant 1} \left(-\infty, \left(n - \frac{2}{3}\right) a_n\right].$$

Given x in $\mathbf{R}$, compute a positive integer $N > x$. If $a_N = 1$, then $x \in (-\infty, N] \subset S$. If $a_N = 0$, then there exists $n \leqslant N-1$ such that $a_{n+1} = 1 - a_n$. Either $x \in (-\infty, n] = S$ or else $x \in [n - \frac{1}{3}, \infty) = S'$. Thus $\mathbf{R} = S \cup S'$;

similarly, $\mathbf{R} = T \cup T'$. Also, $\rho(y,z) \geqslant \frac{1}{3}$ for all $y \in S'$ and $z \in T'$; so $S' \bowtie T'$. Thus (S,T) is a neat cover of $\mathbf{R}$. Suppose that $[0,\infty)$ is neatly located in $\mathbf{R}$. Then either $[0,\infty) \subset S$ or $[0,\infty) \cap T$ is inhabited. In the first case, $a_n = 1$ for all n. In the second, there exists a positive integer ν such that $[0,\nu) \cap T$ is inhabited; whence $a_n = 0$ for some $n \leqslant \nu + 1$. This proves *only if.*

To prove *if*, assume **LPO** and consider any neat cover (S,T) of $\mathbf{R}$. For each positive integer n, the totally bounded interval $[0,n]$ is neatly located, by Proposition 3.10.4. We can therefore construct a binary sequence $(a_n)_{n\geqslant 1}$ such that

$$a_n = 0 \Rightarrow [0,n] \subset S,$$
$$a_n = 1 \Rightarrow [0,n] \cap T \neq \varnothing.$$

By **LPO**, either $a_n = 0$ for all n and therefore $[0,\infty) \subset S$, or else there exists n such that $[0,n] \cap T$, and therefore $[0,\infty) \cap T$, is inhabited. Hence $[0,\infty)$ is neatly located. ∎

One nice feature of neat locatedness is that it is preserved under strong continuity.

Proposition 3.10.6 *Let $f : X \to Y$ be a strongly continuous mapping between symmetric pre-apartness spaces, and let A be a neatly located subset of X. Then $f(A)$ is a neatly located subset of Y.*

Proof. Consider any neat cover (S,T) of Y. There exist subsets S',T' of Y such that $S' \bowtie T'$ and $Y = S \cup S' = T \cup T'$. Then $f^{-1}(S') \bowtie f^{-1}(T')$, by strong continuity, and

$$X = f^{-1}(S) \cup f^{-1}(S') = f^{-1}(T) \cup f^{-1}(T').$$

So $(f^{-1}(S), f^{-1}(T))$ is a neat cover of X. Hence either $A \subset f^{-1}(S)$ or $A \cap f^{-1}(T)$ is inhabited; from which we see that either $f(A) \subset S$ or $f(A) \cap T$ is inhabited. ∎

Corollary 3.10.7 *Let f be a strongly continuous mapping of a symmetric pre-apartness space X into a metric space Y, and let A be a neatly located subset of X. Then $f(A)$ is located in Y.*

Proof. This follows from Propositions 3.10.6 and 3.10.3. ∎

We now derive a number of lemmas that will lead us to Diener's formulation of the notion of neat compactness.

Lemma 3.10.8 *Let $(X,\mathcal{U})$ be a separable uniform space, let $(x_n)_{n\geqslant 1}$ be a dense sequence in X, and let $(V_1,\dots,V_4)$ be a 4-chain of symmetric entourages of X. Let $(a_k)_{k\geqslant 1}$ be an increasing binary sequence with $a_1 = 0$, and*

$(n_k)_{k\geqslant 1}$ an increasing sequence of positive integers such that for each $k > 1$, if $a_k = 0$, then $n_k > n_{k-1}$. Define

$$S \equiv \bigcup \left\{ \bigcup_{n=1}^{n_k} V_1[x_n] : a_k = 0 \right\}, \quad T \equiv \bigcap \left\{ \bigcap_{n=1}^{n_k} \neg V_4[x_n] : a_k = 0 \right\}.$$

Then (S, T) is a neat cover of X.

Proof. Define

$$S' \equiv \bigcap \left\{ \bigcap_{n=1}^{n_k} \neg V_2[x_n] : a_k = 0 \right\},$$

$$T' \equiv \bigcup \left\{ \bigcup_{n=1}^{n_k} V_3[x_n] : a_k = 0 \right\}.$$

Using Lemma 3.10.1, we see that $S' \bowtie T'$. Consider any $x \in X$. Since $(x_n)_{n\geqslant 1}$ is dense in X, there exists k such that $x_k \in V_1[x]$ and therefore, by symmetry, $x \in V_1[x_k]$. If $a_k = 0$, then $n_k \geqslant k$ and so $x \in S$. If $a_k = 1$, then there exists a unique $j \leqslant k$ such that $a_j = 1 - a_{j-1}$. In that case, since $X = V_1[x_n] \cup \neg V_2[x_n]$ for each n, either there exists $i \leqslant j - 1$ such that

$$x \in \bigcup_{n=1}^{n_i} V_1[x_n] \subset S,$$

or else

$$x \in \bigcap_{i=1}^{j-1} \bigcap_{n=1}^{n_i} \neg V_2[x_n] = S'.$$

It now follows that $X = S \cup S'$. A similar argument shows that $X = T \cup T'$. Hence (S, T) is a neat cover of X. ∎

Lemma 3.10.9 Let $(X, \mathcal{U})$ be a separable and neatly located uniform space, $(x_n)_{n\geqslant 1}$ a dense sequence in $(X, \mathcal{U})$, and (V_1, V_2, V_3, V_4) a 4-chain of symmetric entourages of X. Let $(a_k)_{k\geqslant 1}$ be an increasing binary sequence, and $(n_k)_{k\geqslant 1}$ an increasing sequence of positive integers such that for each $k > 1$, if $a_k = 0$, then $n_k > n_{k-1}$ and

$$x_{n_k} \in \bigcap_{n=1}^{n_{k-1}} \neg V_1[x_n]. \tag{3.56}$$

Then either $a_k = 0$ for all k or else there exists K such that $a_K = 1$.

Proof. We may assume that $a_2 = 0$. Setting

$$S = \bigcup \left\{ \bigcup_{n=1}^{n_i} V_1[x_n] : a_{i+1} = 0 \right\}, \quad T = \bigcap \left\{ \bigcap_{n=1}^{n_i} \neg V_4[x_n] : a_{i+1} = 0 \right\},$$

we see from Lemma 3.10.8 that (S,T) is a neat cover of X; whence either $X = S$ or T is inhabited. In the first case, if there exists k such that $a_{k+1} = 1 - a_k$, then (3.56) holds, by our hypothesis, and

$$x_{n_k} \in S = \bigcup_{n=1}^{n_{k-1}} V_1[x_n],$$

which is plainly absurd; hence $a_k = 0$ for all k. In the second case, pick y in T. Since $(x_n)_{n \geqslant 1}$ is dense in X, there exists K such that $x_K \in V_4[y]$; whence $y \in V_4^{-1}[x_K] = V_4[x_K]$. If $a_K = 0$, then $n_K \geqslant K$ and

$$y \in T \subset \bigcap_{n=1}^{n_K} \neg V_4[x_n] \subset \neg V_4[x_K],$$

a contradiction from which we conclude that $a_K = 1$. ∎

Lemma 3.10.10 *Let $(X,\mathcal{U})$ be a separable, neatly located uniform space, $(x_n)_{n \geqslant 1}$ a dense sequence in X, and $(U_1, \ldots, U_5)$ a 5-chain of symmetric entourages of X. Then there exist an increasing binary sequence $(\lambda_k)_{k \geqslant 1}$ and an increasing sequence $(n_k)_{k \geqslant 1}$ of positive integers such that $\lambda_1 = 0$, $n_1 = 1$, and for each $k > 1$,*

(i) *if $\lambda_k = 0$, then $n_k > n_{k-1}$ and $x_{n_k} \in \bigcap_{n=1}^{n_{k-1}} \neg U_5[x_n]$;*

(ii) *if $\lambda_k = 1$, then $n_k = n_{k-1}$ and $X = \bigcup_{n=1}^{n_{k-1}} U_1[x_n]$.*

Proof. For each n define

$$S_n \equiv \bigcup_{i=1}^{n} U_1[x_i], \quad T_n \equiv \bigcap_{i=1}^{n} \neg U_4[x_i]$$

and

$$S_n' \equiv \bigcap_{i=1}^{n} \neg U_2[x_i], \quad T_n' \equiv \bigcup_{i=1}^{n} U_3[x_i].$$

It is straightforward to show that

$$X = S_n \cup S_n' = T_n \cup T_n'.$$

In view of this and Lemma 3.10.1, we see that (S_n, T_n) is a neat cover of X. Setting $\lambda_1 = 0$ and $n_1 = 1$, suppose we have constructed λ_k, n_k with the applicable properties. If $\lambda_k = 1$, we set $\lambda_{k+1} = 1$ and $n_{k+1} = n_k$. If $\lambda_k = 0$, then, since X is neatly located in itself, either $X = S_{n_k}$, when we set $\lambda_{k+1} = 1$ and $n_{k+1} = n_k$; or else there exists

$$y \in T_{n_k} = \neg \bigcup_{i=1}^{n_k} U_4[x_i]. \tag{3.57}$$

In the latter case, since $(x_n)_{n \geqslant 1}$ is dense in X and U_5 is symmetric, we can pick n_{k+1} such that $y \in U_5 \left[x_{n_{k+1}} \right]$; then $y \in U_4 \left[x_{n_{k+1}} \right]$, so, in view of (3.57), $n_{k+1} > n_k$. If there exists $j \leqslant n_k$ such that $x_{n_{k+1}} \in U_5[x_j]$, then $(x_j, y) \in U_5 \circ U_5 \subset U_4$, so $y \in \bigcup_{n=1}^{n_k} U_4[x_n]$, a contradiction. Hence $x_{n_{k+1}} \in \bigcap_{n=1}^{n_k} \neg U_5[x_n]$; setting $\lambda_{k+1} = 0$ completes the inductive construction of the sequences $(\lambda_k)_{k \geqslant 1}$ and $(n_k)_{k \geqslant 1}$. ∎

Lemma 3.10.11 *Under the hypotheses of* Lemma 3.10.10, *either there exists N such that $X = \bigcup_{n=1}^N U_1[x_n]$ or else the following holds:*

(*)   *There exists a strictly increasing sequence $(n_k)_{k \geqslant 1}$ of positive integers such that $x_{n_k} \in \bigcap_{n=1}^{n_{k-1}} \neg U_5[x_n]$ for each k.*

In the latter event, **LPO** *obtains.*

Proof. Construct a binary sequence $(\lambda_n)_{n \geqslant 1}$ and an increasing sequence $(n_k)_{k \geqslant 1}$ of positive integers as in Lemma 3.10.10. Let $(V_1, \ldots, V_4)$ be a 4-chain of symmetric entourages with $V_1 = U_5$. Applying Lemma 3.10.9 with $a_k = \lambda_k$, we see that either there exists K with $\lambda_\kappa = 1 - \lambda_{K-1}$, when we have $X = \bigcup_{n=1}^{n_K - 1} U_1[x_n]$; or else $\lambda_k = 0$ for all k. In the latter case, $(n_k)_{k \geqslant 1}$ is a strictly increasing sequence of positive integers such that $x_{n_k} \in \bigcap_{n=1}^{n_{k-1}} \neg U_5[x_n]$ for each k; moreover, the hypotheses of Lemma 3.10.9 hold for any increasing binary sequence $(a_k)_{k \geqslant 1}$, so **LPO** holds. ∎

A subset A of a pre-apartness space X is said to be **neatly compact** if it is neatly located and satisfies the following bizarre condition:

> If **LPO** *holds, then there does* not *exist an ascending sequence $V_1 \subset V_2 \subset \cdots$ of open subsets of X such that $A \subset \bigcup_{n \geqslant 1} V_n$ and $A \cap \neg V_n$ is inhabited for each n.*

This condition asserts, in a strong manner, that under **LPO** there is no countable open cover of A that contains a finite subcover.

Proposition 3.10.12 *Every separable, neatly compact uniform space is totally bounded.*

Proof. Let $(X, \mathcal{U})$ be a separable, neatly compact uniform space, and $(x_n)_{n \geqslant 1}$ a dense sequence in X. Given a symmetric open entourage U_1 of X, construct a 5-chain $(U_1, \ldots, U_5)$ of symmetric open entourages. By Lemma 3.10.11, either there exists N such that

$$X = \bigcup_{n=1}^N U_1[x_n] \tag{3.58}$$

or else, as we may suppose, **LPO** and conclusion (*) of that lemma both hold. Writing

$$V_k \equiv \bigcup_{n=1}^{n_k} U_5[x_n],$$

we see that V_k is open in X (by Lemma 3.2.14), $V_k \subset V_{k+1}$, and $x_{n_{k+1}} \in \neg V_k$. Since U_5 is symmetric and $(x_n)_{n \geqslant 1}$ is dense in X, for each $x \in X$ there exists i such that $x \in U_5[x_i]$; picking k with $n_k > i$, we have $x \in V_k$. Hence $X = \bigcup_{n \geqslant 1} V_n$. Since X is neatly compact, we have arrived at a contradiction, from which we conclude that (3.58) must obtain. Thus, since $U_1 \in \mathcal{U}$ is arbitrary, X is totally bounded. ∎

We shall improve on the following partial converse to Proposition 3.10.12 shortly.

Proposition 3.10.13 *Let* $(X, \mathcal{U})$ *be a uniformly complete, totally bounded uniform space with a countable base of entourages. Then* $(X, \bowtie_{\mathcal{U}})$ *is neatly compact.*

Proof. By Proposition 3.10.4, X is neatly located. Let $(U_n)_{n \geqslant 1}$ be a countable base of entourages of X; in view of Lemma 3.6.7, we may assume that for each n, U_n is symmetric and (U_n, U_{n+1}) is a 2-chain. Suppose that **LPO** holds, and that there exist open sets $V_1 \subset V_2 \subset \cdots$, with union X, such that $\neg V_n$ is inhabited for each n. Pick x_1 in X and then n_1 such that $x_1 \in V_{n_1}$. Using countable choice, construct a strictly increasing sequence $(n_k)_{k \geqslant 1}$ of positive integers, and a sequence $(x_k)_{k \geqslant 1}$ in X, such that $x_{k+1} \in V_{n_{k+1}} \cap \neg V_{n_k}$ for each k. Replacing V_k by V_{n_k}, we may assume that $x_{k+1} \in V_{k+1} \cap \neg V_k$. Setting $x_{0,k} \equiv x_k$, we construct sequences $(x_{n,k})_{k \geqslant 1}$ $(n \geqslant 0)$ in X such that the following hold for each $n \geqslant 1$:

(a) $(x_{n,k})_{k \geqslant 1}$ is a subsequence of $(x_{n-1,k})_{k \geqslant 1}$;

(b) $(x_{n,j}, x_{n,k}) \in U_n$ for all j, k.

Having constructed the sequence $(x_{n-1,k})_{k \geqslant 1}$ with the applicable properties, we proceed as follows. Since X is totally bounded, there exist finitely many points $\xi_1, \ldots, \xi_m$ of X such that $X = \bigcup_{i=1}^{m} U_{n+1}[\xi_i]$. By **LPO** and countable choice, there exists $i \leqslant m$ such that $U_{n+1}[\xi_i]$ contains all terms of a subsequence $(x_{n,k})_{k \geqslant 1}$ of $(x_{n-1,k})_{k \geqslant 1}$. Then

$$(x_{n,j}, x_{n,k}) \in U_{n+1}^{-1} \circ U_{n+1} \subset U_n$$

for all j, k. This completes the inductive construction.

Now set $y_n \equiv x_{n,n}$ for each n. If $i \geqslant j \geqslant n$, then it follows from (a) and (b) that $(y_i, y_j) \in U_n$. Hence $(y_n)_{n \geqslant 1}$ is a uniformly Cauchy sequence in X and so, by the uniform completeness of X, converges to a limit $y \in X$. Pick

N such that $y \in V_N$. Since V_N is open and $U_{n+1} \subset U_n$ for each n, there exists $\nu > N$ such that $U_\nu[y] \subset V_N$. For all sufficiently large $n > N$ we have $y_n \in U_\nu[y] \subset V_N$. But, as a simple induction argument shows, $y_n = x_k$ for some $k \geqslant n$, so

$$y_n = x_k \in \neg V_{k-1} \subset \neg V_{n-1} \subset \neg V_N.$$

This contradiction shows that the bizarre condition in the definition of *neatly compact* holds for X. ∎

We now introduce neat notions of Cauchyness, convergence, and completeness that will lead us to a neat equivalent of uniform compactness for a uniform space with a countable base of entourages.

Let $(x_i)_{i \in I}$ be a net in a pre-apartness space X. We say that $(x_i)_{i \in I}$ is a **neatly Cauchy net** if the following condition holds:

If (S_j, T_j) are neat covers of X for $1 \leqslant j \leqslant m$, then there exists i_0 such that

> ▷ either $x_{i_0} \in T_k$ for each $k \leqslant m$,
> ▷ or else there exists $k \leqslant m$ such that $x_i \in S_k$ for all $i \succcurlyeq i_0$.

We say that $(x_i)_{i \in I}$ **converges neatly**, or is **neatly convergent**, to $x \in X$ if the following property holds:

If (S_j, T_j) are neat covers of X for $1 \leqslant j \leqslant m$, then

> ▶ either $x \in T_k$ for each $k \leqslant m$,
> ▶ or else there exist $k \leqslant m$ and i_0 such that $x \in S_k$ and $x_i \in S_k$ for all $i \succcurlyeq i_0$.

We say that X is **neatly complete** if every neatly Cauchy sequence in X converges neatly to a limit in X.

Lemma 3.10.14 *Every uniformly Cauchy net in a uniform apartness space is neatly Cauchy.*

Proof. Let $(X, \mathcal{U})$ be a uniform space, and $(x_i)_{i \in I}$ a uniformly Cauchy net in X. For $1 \leqslant j \leqslant m$, let (S_j, T_j) be a neat cover of X. For each $j \leqslant m$ pick S_j', T_j' such that

$$S_j' \bowtie T_j' \text{ and } X = S_j \cup S_j' = T_j \cup T_j'. \tag{3.59}$$

Then pick $U_j \in \mathcal{U}$ such that $S_j' \times T_j' \subset \neg U_j$. Since $U \equiv U_1 \cap \cdots \cap U_m \in \mathcal{U}$, we can choose i_0 such that $(x_i, x_{i'}) \in U$ for all $i, i' \succcurlyeq i_0$. Either $x_{i_0} \in T_j$ for each $j \leqslant m$, or else there exists $k \leqslant m$ such that $x_{i_0} \in T_k'$. In the latter case, for each $i \succcurlyeq i_0$ we have $(x_i, x_{i_0}) \in U \subset U_k$, so $x_i \notin S_k'$ and therefore $x_i \in S_k$. We now see that $(x_i)_{i \in I}$ is a neatly Cauchy net. ∎

Lemma 3.10.15 *Every neatly Cauchy net in a totally bounded uniform space is uniformly Cauchy.*

Proof. Let $(X, \mathcal{U})$ be a totally bounded uniform space, and $(x_i)_{i \in I}$ a neatly Cauchy net in X. Given an entourage U_1, construct a 5-chain $(U_1, \ldots, U_5)$ of entourages of X, with U_j symmetric for $2 \leqslant j \leqslant 5$. Let $\{\xi_1, \ldots, \xi_m\}$ be a finitely enumerable U_5-approximation to X. We see from Lemma 3.10.2 that for each $j \leqslant m$, the pair $(U_2[\xi_j], \neg U_5[\xi_j])$ is a neat cover of X. By the definition of *neatly Cauchy*, there exists i_0 such that

> $\triangleright$ either $x_{i_0} \in \neg U_5[\xi_j]$ for each $j \leqslant m$, which is absurd in view of our choice of the points ξ_j,

> $\triangleright$ or else, as must be the case, there exists $k \leqslant m$ such that $x_i \in U_2[\xi_k]$ for all $i \succcurlyeq i_0$.

Then for all $i, i' \succcurlyeq i_0$ we have $(x_i, x_{i'}) \in U_2^{-1} \circ U_2 \subset U_1$. Since $U_1 \in \mathcal{U}$ is arbitrary, we conclude that $(x_i)_{i \in I}$ is a uniformly Cauchy sequence. ∎

Proposition 3.10.16 *In a uniform apartness space, neat convergence, convergence in the uniform topology, and (apartness) convergence coincide.*

Proof. Let $(X, \mathcal{U})$ be a uniform apartness space, and $(x_i)_{i \in I}$ a net in X. Suppose first that $(x_i)_{i \in I}$ converges neatly to $x_\infty \in X$. Let $(U_1, \ldots, U_4)$ be a 4-chain of symmetric entourages of U. By Lemma 3.10.2, $(U_1[x_\infty], \neg U_4[x_\infty])$ is a neat cover of X. Hence either $x_\infty \in \neg U_4[x_\infty]$, which is absurd, or else, as must be the case, there exists $i_0 \in I$ such that $x_i \in U_1[x_\infty]$ for all $i \succcurlyeq i_0$. Since $U_1 \in \mathcal{U}$ is arbitrary, it follows that $(x_i)_{i \in I}$ is $\tau_{\mathcal{U}}$-convergent to x_∞.

Now suppose that $(x_i)_{i \in I}$ is $\tau_{\mathcal{U}}$-convergent to $x_\infty \in X$. For $1 \leqslant j \leqslant m$, let (S_j, T_j) be a neat cover of X. For each $j \leqslant m$ pick S_j', T_j' such that (3.59) holds, and then a symmetric $U_j \in \mathcal{U}$ with $S_j' \times T_j' \subset \neg U_j$. Since $U \equiv U_1 \cap \cdots \cap U_m \in \mathcal{U}$, we can choose i_0 such that $x_i \in U[x_\infty]$ for all $i \succcurlyeq i_0$. Either $x_\infty \in T_j$ for each $j \leqslant m$ or else there exists $k \leqslant m$ with $x_\infty \in T_k'$. In the latter case, for each $i \succcurlyeq i_0$ we have

$$(x_i, x_\infty) \in U^{-1} = U \subset U_k,$$

so $x_i \notin S_k'$ and therefore $x_i \in S_k$; likewise, $x_\infty \in S_k$. It now follows that $(x_i)_{i \in I}$ converges neatly to x_∞.

To complete the proof, we observe that, in view of Corollary 3.2.20, $\bowtie_{\mathcal{U}}$-convergence is equivalent to $\tau_{\mathcal{U}}$-convergence. ∎

Proposition 3.10.17 *Let $(X, \mathcal{U})$ be a uniform space. If X is neatly complete, then it is uniformly complete. If X is totally bounded and uniformly complete, then it is neatly complete.*

Proof. Suppose that X is neatly complete, and let $(x_i)_{i \in I}$ be a uniformly Cauchy net in X. Lemma 3.10.14 shows that $(x_i)_{i \in I}$ is neatly Cauchy; whence it converges neatly to a limit $x_\infty \in X$. It follows from Proposition 3.10.16 that $(x_i)_{i \in I}$ converges uniformly to x_∞.

Now suppose that X is both totally bounded and uniformly complete, and consider any neatly Cauchy net $(x_i)_{i \in I}$ in X. Lemma 3.10.15 shows that $(x_i)_{i \in I}$ is uniformly Cauchy; it therefore converges uniformly to an element x_∞ of X. It follows from Proposition 3.10.16 that it converges neatly to x_∞. ∎

Putting all the foregoing together, we can show that neat compactness and neat completeness together form a good extension of the notion of compactness from uniform spaces (at least those that have a countable base of entourages) to general pre-apartness spaces.

Theorem 3.10.18 *The following are equivalent for a uniform space X with a countable base of entourages:*

(i) *X is totally bounded and uniformly complete.*

(ii) *X is separable, neatly compact, and neatly complete.*

Proof. Suppose that X is totally bounded and uniformly complete. By Lemma 3.3.20, X is separable. Proposition 3.10.13 shows that X is neatly compact, Proposition 3.10.17 that it is neatly complete. Hence (i) implies (ii).

Suppose, conversely, that (ii) holds. Then, by Proposition 3.10.12, X is totally bounded. On the other hand, Proposition 3.10.17 shows that X is uniformly complete. Hence (ii) implies (i). ∎

We end with some attractive features of the neat compactness property.

Proposition 3.10.19 *Let X be a neatly compact pre-apartness space, Y a symmetric pre-apartness space, and f a strongly continuous, topologically continuous mapping of X into Y. Then $f(X)$ is neatly compact.*

Proof. Proposition 3.10.6 shows that $f(X)$ is neatly located. Suppose that **LPO** holds and that there exist open subsets $V_1 \subset V_2 \subset \cdots$ of Y such that

$$f(X) = \bigcup_{n \geqslant 1} (f(X) \cap V_n)$$

and $f(X) \cap \neg V_n$ is inhabited for each n. Then, since f is topologically continuous, $f^{-1}(V_n)$ is open. Moreover,

$$f^{-1}(V_1) \subset f^{-1}(V_2) \subset \cdots, \quad X = \bigcup_{n \geqslant 1} f^{-1}(V_n),$$

and $X \cap \neg f^{-1}(V_n)$ is inhabited for each n. Since X is neatly compact, this is impossible. Hence $f(X)$ is neatly compact. ∎

Corollary 3.10.20 *Let X be a neatly compact pre-apartness space, Y a uniform space, and f a strongly continuous mapping of X into Y. Then $f(X)$ is neatly compact.*

Proof. Proposition 3.2.26 shows that Y satisfies **B7** and hence has the weak nested neighbourhoods property. We now see from Proposition 2.3.7 that f is topologically continuous. Reference to Proposition 3.10.19 completes the proof. ∎

Corollary 3.10.21 *Let X be a neatly compact pre-apartness space, and f a strongly continuous mapping of X into $\mathbf{R}$. Then $f(X)$ is neatly compact, and $\sup f(X)$ exists.*

Proof. The preceding corollary shows that $f(X)$ is neatly compact; whence it is located in the separable space $\mathbf{R}$, by Corollary 3.10.7, and is therefore separable (see page 16). We now see from Proposition 3.10.12 that $f(X)$ is totally bounded; whence $\sup f(X)$ exists. ∎

> Now this is not the end. It is not even the beginning of the end. But it is, perhaps, the end of the beginning.
>
> Winston Churchill

Notes on Chapter 3

The axioms we have used for set-set apartness have evolved over the years since we first studied that notion, so they differ in some respects from those found in several earlier papers, such as [27] and [25].

In the setting of a set-set apartness space X, we could make the associated point-set apartness symmetric by adding the axiom

$$A \bowtie \{x\} \Rightarrow x \in -A.$$

We would then have

$$A \bowtie B \Rightarrow B \subset -A,$$

a weak form of symmetry of the set-set apartness. Whether this observation has any practical or theoretical value remains to be explored.

The conclusion of Proposition 3.1.8, which reappears as condition (iv) of Proposition 3.7.12, is always satisfied classically, since in that setting we have $\neg S \cup \neg T = \neg(S \cap T)$. Constructively, although $\neg S \cup \neg T \subset \neg(S \cap T)$, we cannot expect to prove the reverse inclusion in general.

For the classical theory of the apartness corresponding to the proximity δ defined on a topological space by

$$A \, \delta \, B \Leftrightarrow \overline{A} \cap \overline{B} \neq \varnothing$$

see [75] (in particular, Theorem (2.11) and pages 71–72).

Adaptations of the material on the left pre-apartness on a topological group G show that

$$S \bowtie T \Leftrightarrow \exists_{U \in \mathfrak{N}(e)} \left(ST^{-1} \subset \sim U \right)$$

defines a set-set pre-apartness (which we naturally call the **right pre-apartness**) on G whose corresponding point-set apartness is the same as that induced by the given topology on G. Moreover, if G is decomposable and has both a tight inequality and the reverse Kolmogorov property, then, in analogy with Proposition 3.2.35, its right pre-apartness is an apartness and coincides with the apartness associated with a natural uniform structure.

Good references for the classical theory of uniform spaces are [13, 42, 82]. Our current axiom **U2** differs from that in earlier papers, where it took the form

> For all $x, y \in X$, $x \neq y$ if and only if there exists $U \in \mathcal{U}$ such that $(x, y) \in \neg U$.

In that case, we required the underlying set X to be equipped with an inequality relation from the outset. However, this approach did not admit of a proof that the inequality on X is tight. On reflection, it became clear that the right approach was not to require an a priori inequality relation on X and the above version of the axiom, but to require that the intersection of $\mathcal{U}$ be the diagonal of $X \times X$ and then to define the inequality on X as we did on page 78. Note, incidentally, that this procedure, together with axiom **U4**, forces a uniform space to be Hausdorff, something that is not required in most classical developments of the theory.

Another substantial improvement in the theory is produced by the replacement of $\sim$ (used in our earlier papers) by $\neg$ in the definition of n-*chain*.

The context should make it clear whether such notations as U^{-1} and U^3 refer to subsets of a topological group or to subsets of the Cartesian product of a quasi-uniform space with itself.

In Lemma 3.2.12 we proved that

$$\sim (U[x]) \subset (\sim U)[x] \tag{3.60}$$

for any *open* entourage U of a uniform space $(X, \mathcal{U})$ and any $x \in X$. Now let $\mathcal{U}$ be a quasi-uniformity on X, and, using Proposition 3.2.6, construct $V \in \mathcal{U}$ such that $V^2 \subset U$ and $X \times X = U \cup \sim V$. Taking $S = \sim (U[x])$, since $U[x] \subset \sim S$, we see from the last part of Lemma 3.2.9 that $\{x\} \times \sim (U[x]) \subset \sim V$; whence $\sim (U[x]) \subset (\sim V)[x]$. Thus we obtain a weak constructive substitute for (3.60).

Here is a classical proof that the Efremovič property implies **B7**. Given that $S \bowtie T$ and applying **EF**, construct E such that $S \bowtie \neg E$ and $E \bowtie T$. Consider any $x \in X$. Either $x \bowtie S$ or near(x, S). In the first case take $R \equiv S$; then, by *ex falso quodlibet*, if $S - R$ is inhabited, we have $\neg R \bowtie T$. In the case where near(x, S) we have $x \bowtie \neg E$ and therefore $x \in --\neg E$; also, $\neg \neg E = E \bowtie T$, so it remains to take $R \equiv \neg E$.

A symmetric $\mathbf{T}_1$ pre-apartness space X with the property $\mathbf{B7}$ is Hausdorff. For if $x \neq y$ in X, then (X being $\mathbf{T}_1$) $x \bowtie \{y\}$; so, by $\mathbf{B7}$, there exists $U \subset X$ such that $x \in -U$ and $\neg U \bowtie \{y\}$. Hence $y \bowtie \neg U$, so $y \in -\neg U$. Since $-\neg U \subset {\sim}\neg U \subset {\sim} - U$, we have the desired conclusion.

In the classical theory of uniform spaces, and in the constructive approach suggested by Errett Bishop in his 1967 monograph ([9], page 110, Exercises 17–21), an important role is played by **pseudometrics**: that is, mappings $\rho : X \to \mathbf{R}$ such that for all $x,y,z \in X$,

- if $x = y$, then $\rho(x,y) = 0$;

- $\rho(x,y) = \rho(y,x)$; and

- $\rho(x,z) \leqslant \rho(x,y) + \rho(y,z)$.

Every family $(\rho_i)_{i \in I}$ of pseudometrics on an inhabited set X induces a uniform structure $\mathcal{U}$ in which sets of the form

$$\{(x,y) \in X \times X : \forall_{i \in F} \left(\rho_i(x,y) < \varepsilon \right)\},$$

with $\varepsilon > 0$ and F an inhabited, finitely enumerable subset of I, form a base of entourages. Classically, every uniform structure on X is induced by a family of pseudometrics on X; moreover, a given uniform structure is induced by a *single* pseudometric (which, for a Hausdorff uniform space, is a metric) if and only if it has a countable base of entourages. A careful analysis of the classical proofs in [62] (pages 184–190), [82] (pages 128–132), or [13] (Part 2, pages 142–144), suggests that, although a metrisable uniform space clearly has a countable base of entourages, the converse is exceedingly unlikely to be provable constructively.

Another example of a pre-uniform, but not uniform, space is the space $\mathbf{R}_d$: that is, the set $\mathbf{R}$ taken with the denial inequality. It is straightforward to show that the supersets of the diagonal Δ of $\mathbf{R} \times \mathbf{R}$ form a pre-uniform structure $\mathcal{U}_d$ on $\mathbf{R}_d$. However, this pre-uniform structure does not satisfy axiom $\mathbf{B5}$: if it did, then there would exist $V \in \mathcal{U}_d$ such that $\mathbf{R} \times \mathbf{R} = \Delta \cup \neg V$ and therefore

$$\forall_{x,y \in \mathbf{R}} \left(x = y \vee \neg \left(x = y \right) \right),$$

which is equivalent to $\mathbf{WLPO}$. Some authors have raised $\mathbf{R}_d$ as an objection to our inclusion of $\mathbf{B5}$ among our axioms for a uniform structure. However, we believe that $\mathbf{R}_d$ is of considerably lesser significance than the metric uniform space $\mathbf{R}$ in constructive analysis and topology, and that the satisfaction of $\mathbf{B5}$ by most of the major examples of pre-uniform spaces justifies our adoption of $\mathbf{B5}$ as an axiom for uniform spaces.

The name *CHN-Cauchy* comes from Cameron, Hocking, and Naimpally [35]. In our earlier papers we used the terms *Cauchy* and *(apartness) complete* instead of the *totally Cauchy* and *totally complete* that we have used above.

For the classical equivalence, in a uniform space, of compactness (in the standard sense: every open cover has a finite subcover) and total boundedness plus completeness see [13] (Chapter 1, §9.1, and Chapter 2, §§4.1 and 4.2) or [82] (Chapter 1, Section 7.1, and Chapter 2, Sections 3.6 and 3.7). The latter reference also has a good discussion of filters and ultrafilters, in Sections 5.1 and 5.7.

There is another classical approach to proving that, for uniform spaces, total completeness implies compactness. Given a totally complete uniform space $(X, \mathcal{U})$, with classical logic we can prove that the set $\mathcal{B}_w$ discussed in Section 3.9 generates the unique totally bounded uniform structure $\mathcal{T}$ that induces the set-set apartness $\bowtie_{\mathcal{U}}$ on X. It follows from Proposition 3.5.16 that $(X, \mathcal{T})$ is uniformly complete; it is therefore compact. Now, the apartness relations $\bowtie_{\mathcal{U}}$ and $\bowtie_{\mathcal{T}}$ coincide, as therefore do their apartness topologies; it follows from Corollary 3.2.20 that $\tau_{\mathcal{U}}$ is compact.

A consequence of the preceding paragraph is that if $(X, \mathcal{U})$ is a complete, but not totally bounded, uniform space, then classically it must contain a $\bowtie_{\mathcal{U}}$-totally Cauchy net that is not uniformly Cauchy. In fact, there must be an ultrafilter on X that is not convergent; the corresponding net is then totally Cauchy but not convergent.

To derive the law of excluded middle from the proposition that every closed subspace of a compact metric space is totally bounded, take X to be the complete, totally bounded, discrete metric space $\{-1, 0, 1\}$, and consider any statement P. Since the terms of a convergent net in X are eventually constant, the subset

$$S \equiv \{0\} \cup \{x : (x = -1 \wedge \neg P) \vee (x = 1 \wedge P)\}$$

of X is (nearly) closed. If it is totally bounded, then we can find a finite set $T \subset \{-1, 0, 1\}$ such that $\rho(s, T) < 1$ for each $s \in S$. If $-1 \in T$, then $\neg P$ holds; if $1 \in T$, then P holds.

It is shown in [22] that if every almost located subset of the Euclidean space $\mathbf{R}^2$ is located, then we can derive the law of excluded middle.

For metric spaces, our notion of *weak-locally totally bounded* is a substantial weakening of that of *locally totally bounded*, as found on page 15. The analogue of Theorem 3.6.14 for locally totally bounded metric spaces is Proposition 2.2.18 of [29].

It is tempting to define the product apartness by saying that two subsets S, T of the product apartness space $X_1 \times X_2$ are apart if there exists $k \in \{1, 2\}$ such that $\mathrm{pr}_k(S) \bowtie \mathrm{pr}_k(T)$. Suppose we adopt that definition. In the product apartness space $\mathbf{R} \times \mathbf{R}$, let C_r denote the square with centre at the origin, sides parallel to the axes, and side-length r. Then the sets $S \equiv C_{1/3}$ and $T \equiv C_1 \sim C_{1/2}$ are bounded away from each other in the product metric on $\mathbf{R} \times \mathbf{R}$; yet they are not apart relative to our suggested alternative notion of product apartness. So that notion is of no use if we want sets that are metrically bounded away from each other to be apart. However, the foregoing sets S, T *are* apart relative to the product apartness as defined in our main text.

The primary reason for restricting our attention to the product of *apartness* spaces is that we need local decomposability (or something of that sort) in the second half of the proof of Proposition 3.7.1. That proposition is used in the proof that the product apartness satisfies axiom **B4**. (If you try to verify this axiom directly, you still need something like local decomposability.)

The example following Corollary 3.7.4 is based on one in Section 4 of [65].

According to the definition given in the article on pages 351–354 of [50], a metric space X is proximally connected if for all $S \subset X$,

$$S \times \neg S \neq \varnothing \Rightarrow S \, \delta \, \neg S, \qquad (3.61)$$

where δ denotes the binary relation of proximity defined, for subsets S,T of X, by

$$S \, \delta \, T \Leftrightarrow \forall_{\varepsilon > 0} \exists_{s \in S} \exists_{t \in T} \, (\rho(s,t) < \varepsilon).$$

However, if the space $[0, 1]$ is proximally connected in this sense, then we can prove the weak law of excluded middle, as follows. Define

$$S \equiv \{0\} \cup \left\{t : 0 \leqslant t \leqslant \tfrac{1}{4} \wedge P\right\} \cup \left\{t : 0 \leqslant t \leqslant \tfrac{3}{4} \wedge \neg P\right\}.$$

Then the pair $(0, 1)$ belongs to $S \times \neg S$. Supposing that $S \, \delta \, \neg S$, we can find $x \in S$ and $y \in \neg S$ with $|x - y| \leqslant 1/4$. Either $y < 3/4$, in which case $\neg\neg P$ holds; or else $y > 1/2$, in which case $x > 1/4$, and therefore $\neg P$ holds. In view of this Brouwerian example, it should be no surprise that in our constructive framework it is a contrapositive of condition (3.61) that provides us with a decent notion of connectedness for pre-apartness spaces.

Various classically equivalent, but constructively inequivalent, concepts of connectedness are discussed in some detail in [15] (pages 80–90).

It is not hard to show that the following property, which is a consequence of (3.49) and which, in view of **B1**, holds classically, would have enabled us to carry through the proof of the last part of Proposition 3.9.3 without (3.49):

$$(S \neq \varnothing \Rightarrow S \bowtie T) \Rightarrow S \bowtie T. \qquad (3.62)$$

Unfortunately, there is no hope of proving (3.62) even in the context of a metric space. To see this, let $(a_n)_{n \geqslant 1}$ be a binary sequence with at most one term equal to 1, and define subsets of the apartness (metric) space $\mathbf{R}$ by

$$S \equiv \left\{\frac{1}{n} : a_n = 1\right\}, \quad T \equiv \{0\}.$$

If there exists n such that $a_n = 1$, then $S = \{1/n\}$ and $S \bowtie T$. But if, without knowing a priori that S is inhabited, we can prove that $S \bowtie T$, then we can find $r > 0$ such that $\rho(s,t) > r$ for all $s \in S$ and all $t \in T$. Then, choosing N such that $1/N < r$, by testing $a_1, \ldots, a_N$ we can show that either there exists $n \leqslant N$ such that $a_n = 1$, or else $a_n = 0$ for all n. Thus property (3.62) entails **LPO**.

Since every uniform apartness space has the Efremovič property, the work in Section 3.9 suggests that we might define an apartness space $(X, \bowtie)$ to be compact if

$\triangleright_a$ it has the Efremovič property and

$\triangleright_b$ $\mathcal{B}_w$, as defined above, strongly generates a uniform structure $\mathcal{U}_w$ that is uniformly complete.

In that case, by Proposition 3.9.13, $\mathcal{U}_w$ is totally bounded. Also, by Corollary 3.9.8, the associated uniform topology $\tau_{\mathcal{U}_w}$ is just $\tau_{\bowtie}$; whence, classically, the topological space $(X, \tau_{\bowtie})$ is compact in any of the usual classically equivalent senses. On the other hand, if $(X, \mathcal{U})$ is any complete, totally bounded uniform space, then by Proposition 3.9.14, $\mathcal{U} = \mathcal{U}_w$, so X is compact in our proposed sense. However, it seems to be hard, if not impossible, to prove, under the exclusion of contradiction arguments, that if a uniform apartness space is compact in the latter sense, then it is both complete and totally bounded (cf. [82], page 142).

Condition $\triangleright_b$ of the previous paragraph can reasonably be weakened to the requirement, not that $\mathcal{B}_w$ generate a complete *uniform* structure, but simply that every net $(x_n)_{n\in\mathfrak{D}}$ in X that satisfies the Cauchy-like condition,

$$\forall_{U\in B_w} \exists_{N\in\mathfrak{D}} \forall_{m,n\succcurlyeq N} \ ((x_m, x_n) \in U),$$

converge to a limit in X. The first author has some preliminary results on compactness notions that use this relaxation of completeness; see [20].

The neat approach to a notion of compactness taken in Section 3.10 is due to Diener [39], who sought a notion of compactness with these two requirements:

▶ For a uniform space, the new notion of compactness should be equivalent to the usual one of completeness plus total boundedness.

▶ Every strongly continuous mapping of a compact pre-apartness space into $\mathbf{R}$ should have a supremum (and an infimum).

The role of **LPO** in Diener's definitions and results is distinctly unusual. In models of **BISH** where **LPO** is false—such as the intuitionistic and the recursive models (see [24])—it is not hard to show (using Lemma 3.10.11 and Proposition 3.10.4) that a separable uniform space is totally bounded if and only if it is neatly located. On the other hand, as we showed in Proposition 3.10.5, the interval $[0,\infty)$ in $\mathbf{R}$ is neatly located if and only if **LPO** holds.

Recently, T. Steinke has investigated other notions of compactness for apartness spaces [86], motivated by the classical result that on a compact Hausdorff space there is a unique proximity—the one corresponding to the apartness defined at (3.10)—whose apartness topology is the given compact one [93] (Theorem 41.1). Also, Bridges has investigated notions of compactness and precompactness in apartness spaces [19, 20].

Among the topics on apartness that we chose to omit from this book are a version of Urysohn's lemma [21] in which an extended type of real numbers is used, and the first steps in a point-free development of a theory of apartness on lattices [31, 17, 18].

POSTLUDE: PATHS TO TOPOLOGY

There are nine and sixty ways of constructing tribal lays,

And every single one of them is right!

Rudyard Kipling

Other than the one we have followed in the foregoing chapters, the main constructive approaches to topology include:

- a theory of neighbourhood spaces;
- the theory of frames/locales;
- formal topology.

The first of these, originally suggested by Bishop [9, 12], has been connected by Ishihara et al. [48, 59] to the theory of *quasi-apartness spaces*—essentially spaces with a relation $\bowtie$ that satisfy axioms **A1–A4** but with the complement replaced by the logical complement in **A2** and **A4**. In particular, the paper [59] shows how to construct an adjunction between the category[1] of neighbourhood spaces and that of quasi-apartness spaces.

Both the theory of frames/locales and formal topology produce *point-free topology*: topology in which the notion of *point* is secondary to that corresponding to *open set*. In the former case, the fundamental structure is a *frame*, defined to be a complete lattice L in which finite meets distribute over arbitrary unions: that is,

$$x \wedge \left(\bigvee_{i \in I} y_i \right) = \bigvee_{i \in I} (x \wedge y_i)$$

---

[1]In this postlude we assume some familiarity with categories, lattice theory, and orderings.

holds for all $x \in L$ and all families $(y_i)_{i \in I}$ of elements of L. Taken with supremum-preserving lattice homomorphisms, frames form a category that generalises the category of topological spaces with continuous functions as morphisms. Since in the latter category it is not the image, but the inverse image, of an open set that is open, it makes sense to consider the dual of the former category; that dual is called the category of *locales*.

We shall not discuss locales further here,[2] since the resulting development of topology, though constructive, is impredicative and there is a purely predicative constructive theory—*formal topology*—that we now discuss briefly. This was introduced in the mid 1980s by Sambin and Martin-Löf [79, 80, 81], in order to provide a theory of topology that was based on the latter's type theory [69, 70, 71], which is one of the main alternatives proposed as a foundation for **BISH**. The fundamental notions of formal topology are

- a preordered set $(A, \leqslant)$ whose elements are called *basic neighbourhoods*, and

- a so-called *covering relation* $\lhd$ between elements of A and subsets of A,

satisfying four simple axioms. Although points do not have the status in formal topology that they have in classical topology or our theory of apartness spaces, one can introduce a notion of *point* as a special type of subset α of A; roughly, a point is identified with the set of all its neighbourhoods. Denoting the collection[3] of points by $\mathrm{Pt}(A)$, to each set U of basic neighbourhoods we associate a collection U^* of points as follows:

$$U^* \equiv \{\alpha \in \mathrm{Pt}(A) : \exists_{a \in U} (a \in \alpha)\}.$$

It turns out that these collections form a topology, in the usual sense, on $\mathrm{Pt}(A)$. Moreover, when we look at the so-called *continuous morphisms*, or *approximable mappings*, between formal topologies $(A, \leqslant, \lhd)$ and $(A', \leqslant', \lhd')$, we see that to each such relation $F \subset A \times B$ there is associated a *point function* $\mathrm{Pt}(F) : \mathrm{Pt}(A) \to \mathrm{Pt}(A')$ defined by

$$\mathrm{Pt}(F)(\alpha) \equiv \{a' \in A' : \exists_{a \in \alpha} (aFa')\}.$$

An important feature of formal topology (and of locale theory) is a notion of compactness analogous to compactness as defined, both classically and intuitionistically, in terms of open covers. By using formal covers, rather than pointwise ones, the formal topologist avoids the need for Brouwer's fan theorem or its classical counterpart, König's lemma, when establishing, for example, the compactness of the interval $[0, 1]$.

---

[2] For more on topology via locales, see the books by Johnstone [60] and Vickers [89].

[3] We use the word 'collection' since in a predicative foundation like Martin-Löf type theory, $\mathrm{Pt}(A)$ may not be a set.

How does apartness arise within the context of a formal topology A? We say that two basic neighbourhoods a, b in A are *apart*, and we write $a \perp b$, if $a \wedge b \lhd \varnothing$. This gives rise to notions of

- inequality/apartness between points α, β: $\alpha \neq \beta$ if and only if $a \perp b$ for some $a \in \alpha$ and $b \in \beta$;

- apartness between a point α and a subset S of A: $\alpha \bowtie S$ if and only if there exists $a \in \alpha$ such that for each $\beta \in S$, there exists $\beta \in \beta$ with $a \perp b$.

It transpires that if the formal topology satisfies a certain *regularity* property, then $X \equiv \mathrm{Pt}(A)$, taken with the foregoing point-point inequality and apartness between points and sets, satisfies our axioms **A1–A5** for an apartness space. Moreover, if $F : A \to A'$ is an approximable mapping between formal topologies, then the associated mapping $\mathrm{Pt}(F) : \mathrm{Pt}(A) \to \mathrm{Pt}(A')$ is continuous in the point-set apartness sense introduced on page 36.

To extend these ideas to apartness between subsets of our formal space A, first define apartness of two sets U, V of basic neighbourhoods by

$$U \perp V \Leftrightarrow \forall_{u \in U} \forall_{v \in V} (u \perp v).$$

For subsets S, T of A, We then define the apartness between subsets S, T of A by

$$S \bowtie T \Leftrightarrow \exists_{U, V \subset A} (S \subset U^*, \ T \subset V^*, \text{ and } U \perp V).$$

When A is regular as a formal topology, the relation satisfies our axioms **B1**, **B3–B5** (and hence **B4**) and this weak form of **B2**:

$$S \bowtie T \Rightarrow S \cap T = \varnothing.$$

(In fact, we do not need the regularity property of A to establish **B1**, the weak form of **B2**, and **B3–B4**.) Furthermore, if $F : A \to A'$ is an approximable mapping between formal topologies, then $\mathrm{Pt}(F) : \mathrm{Pt}(A) \to \mathrm{Pt}(A')$ is strongly continuous.

Thus there is a connection from formal topologies to apartness spaces (or at least, in the set-set case, spaces which differ from apartness ones by having a weaker form of **B2**). For the reverse connection, we need only pass from the apartness topology associated with a given apartness space $(X, \bowtie)$ to the corresponding formal topology.

While on the subject of lattice-based approaches to topology, we would point out that our theory of apartness between sets has been abstracted to one of apartness on frames [17, 18, 31].

[1] O. Aberth: *Computable Analysis*, McGraw-Hill, New York, 1980.

[2] P. Aczel and M. Rathjen: *Notes on Constructive Set Theory*, Report No. 40, Institut Mittag-Leffler, Royal Swedish Academy of Sciences, 2001.

[3] P. Aczel and M. Rathjen: *Constructive Set Theory*, in preparation.

[4] B. Banaschewski, T. Coquand, and G. Sambin (eds): *Second Workshop on Formal Topology* (Venice, 2002), Ann. Pure Appl. Logic **137**, 1–3 (special issue), 2006.

[5] A. Bauer: 'Realizability as the Connection between Computable and Constructive Mathematics', preprint, University of Ljubljana, Slovenia, 2005.

[6] M.J. Beeson: *Foundations of Constructive Mathematics*, Springer Verlag, Heidelberg, Germany, 1985.

[7] H.L. Bentley, H. Herrlich, and M. Hušek: 'The historical development of uniform, proximal, and nearness concepts in topology', in *Handbook of the History of General Topology*, Vol. 2 (C.E. Aull and R. Lowen, eds), 577–629, Kluwer, Amsterdam, The Netherlands, 1998.

[8] J. Berger, H. Ishihara, E. Palmgren, and P.M. Schuster: 'A predicative completion of a uniform space', preprint, University of Munich, Germany, 2010.

[9] E.A. Bishop: *Foundations of Constructive Analysis*, McGraw-Hill, New York, 1967.

[10] E.A. Bishop: 'Schizophrenia in Contemporary Mathematics', in *Errett Bishop, Reflections on Him and His Research*, Contemp. Math. **39**, 1–32, Am. Math. Soc., Providence RI, 1985.

[11] E.A. Bishop: *The Neat Category of Stratified Spaces*, preprint, University of California, San Diego, 1971.

[12] E.A. Bishop and D.S. Bridges: *Constructive Analysis*, Grundlehren der Math. Wissenschaften **279**, Springer Verlag, Heidelberg-Berlin-New York, 1985.

[13] N. Bourbaki: *General Topology* (2 vols: Parts 1 and 2), Addison-Wesley, Reading, Mass., 1966.

[14] D.S. Bridges: 'A constructive Morse theory of sets', in *Mathematical Logic and its Applications* (D. Skordev, ed.), Plenum, New York, 61–79, 1987.

[15] D.S. Bridges: 'A constructive look at the real number line', in *Synthese: Real Numbers, Generalizations of the Reals and Theories of Continua* (P. Ehrlich, ed.), 29–92, Kluwer Academic, Amsterdam, The Netherlands, 1994.

[16] D.S. Bridges: 'Constructive mathematics: a foundation for computable analysis', Theor. Comp. Sci. **219**(1-2), 95–109, 1999.

[17] D.S. Bridges: 'Product a-frames and proximity', Math. Logic Q. **54**(1), 12–25, 2008.

[18] D.S. Bridges: 'Almost new pre-apartness from old', Ann. Pure Appl. Logic, to appear.

[19] D.S. Bridges: 'Precompact apartness spaces', preprint, University of Canterbury, New Zealand, 2011.

[20] D.S. Bridges: 'Some new compactness notions for an apartness space', preprint, University of Canterbury, New Zealand, 2011.

[21] D.S. Bridges and H. Diener: 'A constructive treatment of Urysohn's lemma in an apartness space', Math. Logic Q. **52**(5), 464–469, 2006.

[22] D.S. Bridges, H. Ishihara, R. Mines, F. Richman, P.M. Schuster, and L.S. Vîţă: 'Almost locatedness in uniform spaces', Czech. Math. J. **57**(1), 1–12, 2007.

[23] D.S. Bridges, H. Ishihara, P.M. Schuster, and L.S. Vîţă: 'Strong continuity implies uniform sequential continuity', Arch. Math. Logic **44**(7), 887–895, 2005.

[24] D.S. Bridges and F. Richman: *Varieties of Constructive Mathematics*, London Math. Soc. Lecture Notes **97**, Cambridge Univ. Press, Cambridge, U.K., 1987.

[25] D.S. Bridges, P.M. Schuster, and L.S. Vîță: 'Apartness as a relation between subsets', in *Combinatorics, Computability and Logic* (Proceedings of DMTCS'01, Constanța, Romania, 2–6 July 2001; C.S. Calude, M.J. Dinneen, S. Sburlan, eds.), 203–214, DMTCS Series **17**, Springer Verlag, London, 2001.

[26] D.S. Bridges and L.S. Vîță: 'Characterising near continuity constructively', Math. Logic Q. **47**(4), 535–538, 2001.

[27] D.S. Bridges and L.S. Vîță: 'Cauchy Nets in the Constructive Theory of Apartness Spaces', Scientiae Math. Jpn. **56**, 123–132, 2002.

[28] D.S. Bridges and L.S. Vîță: 'A constructive theory of point-set nearness', in *Topology in Computer Science: Constructivity; Asymmetry and Partiality; Digitization* (Proc. Dagstuhl Seminar 00231, 4-9 June 2000; R. Kopperman, M. Smyth, D. Spreen, eds.), Theor. Comp. Sci. **305**(1–3), 473–489, 2003.

[29] D.S. Bridges and L.S. Vîță: *Techniques of Constructive Analysis,* Universitext, Springer, New York, 2006.

[30] D.S. Bridges and L.S. Vîță: 'Proximal connectedness', Fund. Informaticae **83**(1-2), 25–34, 2008.

[31] D.S. Bridges and L.S. Vîță : 'A constructive theory of apartness on lattices', Scientiae Math. Jpn. **69**(2), 187–206, 2009.

[32] L.E.J. Brouwer: 'Over de onbetrouwbaarheid der logische principes', Tijdschrift voor Wijsbegeerte **2**, 152–158, 1908. English translation in [34], 107–111.

[33] L.E.J. Brouwer: 'Intuitionistische Zerlegung mathematischer Grundbegriffe', Jahresbericht der Deutsche Mathematiker-Vereinigung **33**, 241–256, 1924.

[34] L.E.J. Brouwer: *Collected Works 1* (A. Heyting, ed.), North-Holland, Amsterdam, The Netherlands, 1975.

[35] P. Cameron, J.G. Hocking, and S.A. Naimpally: *Nearness—a better approach to topological continuity and limits,* Amer. Math. Monthly **81**(7), 739–745, 1974.

[36] R.L. Constable et al.: *Implementing Mathematics with the Nuprl Proof Development System,* Prentice-Hall, Englewood Cliffs, New Jersey, 1986.

[37] L.S. Dediu (Vîță) and D.S. Bridges: 'Constructive notes on uniform and locally convex spaces', in *Proceedings of International Symposium FCT '99* (Iași, Romania), Springer Lecture Notes in Computer Science **1684,** 195–203, 1999.

[38] R. Diaconescu: 'Axiom of choice and complementation', Proc. Am. Math. Soc. **51**, 176–178, 1975.

[39] H. Diener: 'Generalising compactness', Math. Logic Q. **51**(1), 49–57, 2008.

[40] M.A.E. Dummett: *Elements of Intuitionism* (Second Edition), Oxford Logic Guides **39**, Clarendon Press, Oxford, 2000.

[41] V.A. Efremovič: 'Infinitesimal spaces', Dokl. Akad. Nauk. USSR **76**, 341–343, 1951 (in Russian).

[42] R. Engelking: *General Topology*, Heldermann Verlag, Berlin, Germany, revised and completed edition, 1989.

[43] H.M. Friedman: 'Set Theoretic Foundations for Constructive Analysis', Ann. Math. **105**(1), 1–28, 1977.

[44] R.I. Goldblatt: *Topoi*, North-Holland, Amsterdam, The Netherlands, 1979.

[45] N. D. Goodman and J. Myhill: 'Choice Implies Excluded Middle', Zeit. Logik und Grundlagen der Math. **24**, 461, 1978.

[46] R.J. Grayson: 'Concepts of general topology in constructive mathematics and in sheaves I', Ann. Math. Logic **20**, 1–41, 1981.

[47] R.J. Grayson: 'Concepts of general topology in constructive mathematics and in sheaves II', Ann. Math. Logic **23**, 55–98, 1982.

[48] R.S. Havea, H. Ishihara, and L.S. Vîţă: 'Separation properties in neighbourhood and quasi-apartness spaces', Math. Logic Q. **54**(1), 58–64, 2008.

[49] S. Hayashi and H. Nakano: *PX: A Computational Logic*, MIT Press, Cambridge, Mass., 1988.

[50] M. Hazewinkel (ed.): *Encyclopaedia of Mathematics*, Vol. 7, Kluwer Academic, Dordrecht, The Netherlands, 1991.

[51] A. Hedin: 'A note on set-presentable apartness spaces', preprint, University of Uppsala, Sweden, 2010.

[52] H. Herrlich: 'On the extendibility of continuous functions', Gen. Topol. Appl. **5**, 213–215, 1974.

[53] A. Heyting: 'Die formalen Regeln der intuitionistischen Logik', Sitzungsber. preuss. Akad. Wiss. Berlin, 42–56, 1930.

[54] A. Heyting: *Mathematische Grundlagenforschung. Intuitionismus. Beweistheorie*, Springer Verlag, Berlin, 1934.

[55] H. Ishihara: 'Continuity and nondiscontinuity in constructive mathematics', J. Symbolic Logic **56**(4), 1349–1354, 1991.

[56] H. Ishihara: 'Markov's principle, Church's thesis, and Lindelöf's theorem', Indag. Math. N.S. **4**(3), 321–325, 1993.

[57] H. Ishihara: 'A constructive version of Banach's inverse mapping theorem', N.Z. J. Math. **23**, 71–75, 1994.

[58] H. Ishihara: 'Two subcategories of apartness spaces', Ann, Pure Appl. Logic, to appear.

[59] H. Ishihara, R. Mines, P.M. Schuster, and L.S. Vîţă: 'Quasi-apartness and neighbourhood spaces', Ann. Pure Appl. Logic **141**, 296–306, 2006.

[60] P.T. Johnstone: *Stone Spaces*, Cambridge Studies in Advanced Mathematics **3**, Cambridge Univ. Press, Cambridge, U.K., 1982.

[61] P.T. Johnstone: 'The point of pointless topology', Bull. Am. Math. Soc. **8**(1), 41–53, 1983.

[62] J.L. Kelley: *General Topology*, van Nostrand, Princeton, New Jersey, 1955; re-published as Graduate Text in Mathematics **27**, Springer Verlag, Heidelberg, Germany, 1975.

[63] A.N. Kolmogorov: 'Deutung der intuitionistischen Logik', Math. Z. **35**, 58–65, 1932.

[64] B.A. Kushner: *Lectures on Constructive Mathematical Analysis*, Am. Math. Soc., Providence RI, 1985.

[65] S. Leader: 'On products of proximity spaces', Math. Ann. **154**, 185–194, 1964.

[66] P. Lietz: *From Constructive Mathematics to Computable Analysis via the Realizability Interpretation*, Dr. rer. nat. thesis, Technische Universität, Darmstadt, Germany, 2004.

[67] M. Mandelkern: 'Connectivity of an interval', Proc. Am. Math. Soc. **54**, 170–172, 1976.

[68] P. Martin-Löf: *Notes on Constructive Mathematics*, Almqvist and Wiksell, Stockholm, Sweden, 1970.

[69] P. Martin-Löf: 'Constructive mathematics and computer programming', in *Logic, Methodology, and Philosophy of Science VI* (J.J. Cohen, J. Loś, H. Pfeiffer, K-P. Podewski, eds), 153–175, North-Holland, Amsterdam, The Netherlands, 1982.

[70] P. Martin-Löf: *Intuitionistic Type Theory*, Notes by G. Sambin, Bibliopolis, Napoli, Italy, 1984.

[71] P. Martin-Löf: 'An intuitionistic theory of types', in *Twenty-five Years of Constructive Type Theory* (G. Sambin, J. Smith, eds), 127–172, Oxford Logic Guides **36**, Clarendon Press, Oxford, 1998.

[72] J. Myhill: 'Constructive set theory', J. Symbolic Logic **40**(3), 1975, 347–382.

[73] L.J. Nachman: 'On a conjecture of Leader', Fund. Math. **LXV**, 153–155, 1969.

[74] S.A. Naimpally: *Proximity Approach to Problems in Topology and Analysis*, Oldenbourg, Munich, Germany, 2009.

[75] S.A. Naimpally and B.D. Warrack: *Proximity Spaces*, Cambridge Tracts in Math. and Math. Phys. **59**, Cambridge Univ. Press, Cambridge, U.K., 1970.

[76] E. Palmgren and P.M. Schuster: 'Apartness and formal topology', N.Z. J. Math. **35**, 77–84, 2006.

[77] F. Richman: 'Constructive mathematics without choice', in *Reuniting the Antipodes—Constructive and Nonstandard Views of the Continuum* (P.M. Schuster, U. Berger, and H. Osswald, eds), 199–205, Kluwer, Synthese Library **306,** 2001.

[78] F. Riesz: 'Stetigkeitsbegriff und abstrakte Mengenlehre', Atti IV Congr. Intern. Mat. Roma **II**, 18–24, 1908.

[79] G. Sambin: 'Intuitionistic formal spaces—a first communication', in *Mathematical Logic and its Applications* (D.G. Skordev, ed.), 187–204, Plenum, New York, 1987.

[80] G. Sambin: 'Some points in formal topology', Theor. Comp. Sci. **305**, 347–408, 2003.

[81] G. Sambin: *The Basic Picture: Structures for Constructive Topology*, Oxford Logic Guides, Clarendon Press, Oxford, to appear.

[82] H. Schubert: *Topology*, Macdonald Technical & Scientific, London, 1968.

[83] P.M. Schuster, L.S. Vîţă, and D.S. Bridges: 'Apartness as a relation between subsets', in *Combinatorics, Computability and Logic* (Proceedings of DMTCS'01, Constanţa, Romania, 2–6 July 2001; C.S. Calude, M.J. Dinneen, S. Sburlan, eds.), 203–214, DMTCS Series **17**, Springer Verlag, London, 2001.

[84] H.A. Schwichtenberg: 'Program extraction in constructive analysis', in *Logicism, Intuitionism, and Formalism—What has become of them?* (S. Lindström, E. Palmgren, K. Segerberg, and V. Stoltenberg-Hansen, eds), 255–275, Synthese Library **341**, Springer Verlag, Berlin, Germany, 2009.

[85] L.A. Steen and J.A. Seebach: *Counterexamples in Topology,* Dover, New York, 1995.

[86] T.A. Steinke: *Constructive Notions of Compactness in Apartness Spaces*, M.Sc. thesis, University of Canterbury, Christchurch, New Zealand, 2011.

[87] A.S. Troelstra: *Intuitionistic General Topology*, Ph.D. Thesis, University of Amsterdam, The Netherlands, 1966.

[88] A.S. Troelstra and D. van Dalen: *Constructivism in Mathematics: An Introduction* (two volumes), North-Holland, Amsterdam, The Netherlands, 1988.

[89] S.J. Vickers, *Topology via Logic*, Cambridge Tracts in Theoretical Computer Science **5,** Cambridge Univ. Press, Cambridge, U.K., 1988.

[90] L.S. Vîţă and D.S. Bridges: 'A constructive theory of point-set nearness', in *Topology in Computer Science: Constructivity; Asymmetry and Partiality; Digitization* (Proc. Dagstuhl Seminar 00231, 4-9 June 2000; R. Kopperman, M. Smyth, D. Spreen, eds.), 473–489, Theor. Comp. Sci. **305**(1–3), 2003.

[91] F.A. Waaldijk: *Constructive Topology*, Ph.D. thesis, University of Nijmegen, The Netherlands, 1996.

[92] K. Weihrauch: *Computable Analysis*, EATCS Texts in Theoretical Computer Science, Springer Verlag, Heidelberg, Germany, 2000.

[93] S. Willard: *General Topology*, Addison-Wesley, Reading, Mass., 1970.